DO ELEPHANTS HAVE KNEES?

DO ELEPHANTS HAVE KNEES?

And Other Stories of Darwinian Origins

CHARLES R. AULT JR.

COMSTOCK PUBLISHING ASSOCIATES
A DIVISION OF
CORNELL UNIVERSITY PRESS
Ithaca & London

First published 2016 by Cornell University Press
Printed in the United States of America

Library of Congress Cataloging-in-Publication Data

Names: Ault, Charles R., Jr., 1950– author.
Title: Do elephants have knees? : and other stories of Darwinian origins / Charles R. Ault Jr.
Description: Ithaca : Comstock Publishing Associates, a division of Cornell University Press, 2016. | Includes bibliographical references and index.
Identifiers: LCCN 2016013024 | ISBN 9781501704673 (cloth : alk. paper)
Subjects: LCSH: Evolution (Biology)—Popular works. | Natural selection—Popular works. | Darwin, Charles, 1809–1882—Anecdotes.
Classification: LCC QH367 .A85 2016 | DDC 576.8—dc23
LC record available at http://lccn.loc.gov/2016013024

Cornell University Press strives to use environmentally responsible suppliers and materials to the fullest extent possible in the publishing of its books. Such materials include vegetable-based, low-VOC inks and acid-free papers that are recycled, totally chlorine-free, or partly composed of nonwood fibers. For further information, visit our website at www.cornellpress.cornell.edu.

Cloth printing 10 9 8 7 6 5 4 3 2 1

For Carlos Calvo Zúñiga

Naturalista

Gracias, Carlos, for sharing your exuberance for life's diversity.

And ever since that day, O Best Beloved, all the Elephants you will ever see, besides all those that you won't, have trunks precisely like the trunk of the 'satiable Elephant's Child.

Rudyard Kipling, "The Elephant's Child"

The whole history of the world, as at present known, although of a length quite incomprehensible by us, will hereafter be recognized as a mere fragment of time, compared with the ages which have elapsed since the first creature, the progenitor of innumerable extinct and living descendants, was created.

Charles Darwin, *On the Origin of Species*

CONTENTS

ACKNOWLEDGMENTS

Writing is both a solitary and a communal task. I wish to express my gratitude to all who have supported this effort.

The arrival of a letter from Kitty Hue-Tsung Liu encouraging me to submit a full draft of the manuscript that would become *Do Elephants Have Knees?* brought great joy. I am grateful for Kitty's championing of the book. Her deft touch led an arduous revision process to a successful conclusion. As a graduate student at Cornell University in science and environmental education in the late 1970s, I enjoyed the privilege of sitting on occasion at the desk that once belonged to Anna Botsford Comstock. Throughout my teaching career, I turned to her *Handbook of Nature Study* often. Thanks to Kitty's good work, and the amazing craftsmanship of her fellow editors Karen Hwa and Amanda Heller, this book carries the Comstock imprint.

Numerous reviewers have mixed praise and criticism, each cycle producing improvements. I have tried to make Darwinian histories inviting and accessible to a wide audience and am deeply indebted to Christine Marie Janis, professor of evolutionary biology at Brown University, for her technical edit of several chapters. Errors that remain are completely my own.

Kristen Hall-Geisler of Indigo Publishing in Portland, Oregon, provided an early and comprehensive developmental edit that put the book on a firm foundation. She worked with me on subsequent revisions, guiding the difficult task of synchronizing whimsical children's stories with serious Darwinian ideas.

Writing rekindled friendship with David Goehring, a classmate from undergraduate days who went on to pursue a distinguished career in publishing. David took a close look at the manuscript and declared it "neither fish nor fowl"—yet intriguing nonetheless. With his feedback, I split the manuscript into two books, one with origins stories and the other with ideas about teaching science that became *Challenging Science Standards: A Skeptical Critique of the Quest for Unity* published in 2015 by Rowman & Littlefield. Fish and fowl went their separate ways.

Dr. Paula Mikkelsen, director of publications at the Paleontological Research Institution/Museum of the Earth in Ithaca, New York, helped me to find my voice as a storyteller rather than an academic writer. She edited two essays for *American Paleontologist* that would expand into chapters of this book.

Several of the essays descend with modification from my teaching at Lewis & Clark College—including an annual field class centered on the fossil localities within the John Day National Monument—and the feedback of my students. I owe special thanks to Sarah Mock, Kate Fisher, and Chris Hedeen for keeping me on target and to Phoebe Skinner for capable editing of the manuscript in its formative stage.

For several years, Jerry Kuykendall invited me to lecture on Darwin Day to his class of aspiring science teachers. Those appearances became the chapters intended to counter the stereotype of Darwin as the eminent Victorian sage.

John van Wyhe's Darwin Online Web project proved utterly indispensable to my research. Without it, this book would not exist. Adam Perkins, manuscript curator at the Cambridge University Library, also provided helpful assistance.

Do Elephants Have Knees? is a book that synthesizes science and history to make the imagery of Darwinism vivid and interesting. I first encountered the idea of using whimsical stories written for children to teach about thinking in the work of philosopher Gareth B. Matthews. His message is timeless and I was most fortunate to stumble upon his writing early in my career.

The literary style is my own, along with much of the interpretation of Darwin's interests in *Flustra*. I am deeply indebted to the biographers and scientists whose research I have drawn upon to tell stories of origins in a novel way. Many researchers have illuminated the evolutionary history of limbs, fish elephants, whales, and birds in astonishing detail, and biographers have laid bare virtually every aspect of Charles Darwin's life. My book celebrates their accomplishments and invites the inner child scientist in everyone to look seriously at their work.

My son Toby, now a member of the Cornell science faculty, launched his career with a year retracing Darwin, mostly in South America, courtesy of a Watson Fellowship. My joining Toby turned me, already hooked on paleontology thanks to John Luther Cisne's tutelage at Cornell, into a true Darwinophile.

I am indebted to Jan Glenn, the book's illustrator, whose drawings resonate with the sense of whimsy framing each chapter. For a short time I had the privilege of helping Jan teach a workshop for teachers on integrating art and science. From that effort grew a field activity at the Oregon Zoo first featured in *Challenging Science Standards: A Skeptical Critique of the Quest for Unity*. I thank Patricia Zline at Rowman & Littlefield for permission to retell that story in the chapter "Out on a Limb."

India Wood, my former fourth-grade student and lifelong friend, has contributed in obvious and unobvious ways to the final work—including guiding me through the Morrison Formation in Colorado.

Carlos Calvo and Carmen Hernandez of the University of Costa Rica and Nancy Aitken, director of the Campanario Biological Field Station, did their best for many years to immerse me and my students in their nation's magnificent natural places. Thanks to them I came to share Darwin's "rapture of rapture," the "chaos of delight" to be found wandering amid "the general luxuriance of the vegetation" in a tropical forest (*Beagle Diary* entry for February 28–29, 1832). Anyone who reads Darwin cannot escape wishing to experience the natural world he encountered as an adventuring young man.

Family support has been essential from beginning to end. As a reader of several sections, my biologist daughter (an aficionado of water mold genomes), Kori, took me to task. Again, errors that remain are my own.

I am grateful to my son Logan, whose companionship and reminders to take bicycling breaks and trips to the gym kept me progressing.

To my wife, Phyllis, I extend limitless appreciation. She has read every version of every chapter multiple times, constantly pushing me to achieve clarity. Most important, she put the wind back in my sails whenever I encountered the writing doldrums.

My thanks to all of you.

DO ELEPHANTS HAVE KNEES?

INTRODUCTION

WONDERFUL RELATIONSHIPS

Stories of Darwinian Origins

> *The relationship, though distant, between the Macrauchenia and the Guanaco, between the Toxodon and the Capybara,—the closer relationship between the many extinct Edentata and the living sloths, ant-eaters, and armadillos, now so eminently characteristic of South American zoology . . . are most interesting facts. This wonderful relationship in the same continent between the dead and the living, will, I do not doubt, hereafter throw more light on the appearance of organic beings on our earth, and their disappearance from it, than any other class of facts.*
>
> Charles Darwin, *Voyage of the* Beagle

In 1859 Darwin completed his masterwork and, as he had foretold in *Voyage of the* Beagle over a decade earlier, indeed cast a bright light on the appearance and disappearance of earth's organic beings. Following a lifelong childlike obsession with scurrying beetles, branching polyps, parasitic brachiopods, and lowly worms, Darwin proclaimed the value of the humblest of organisms. The creation of so much intricate beauty with little apparent purpose had always puzzled him. Now far beyond his inchoate musings as a young scientist, Darwin crafted a daring and mature treatise, *On the Origin of Species*. Its publication indelibly changed the world's understanding of ancestry and descent and laid the groundwork for evolutionary biology.

"Darwinian histories" connect progeny to their forebears, the fossilized dead to the living. The serious themes in Darwin's science often tread the same ground as the whimsical ones found in children's literature. *Do Elephants Have Knees?* examines the thinker and his thought by pairing scientific stories of origins with characters from children's literature. Good storytelling in science reenacts explanation while making these explanations inviting.[1] If curiosity is the starting point, satisfying insight is the goal. The task of *Do Elephants Have Knees?* is to exploit playful stories in order to comprehend Darwin's conception of nature.

The book begins with portraits of Darwin: child, young man, grandfather. Each stage poses the question "What was Darwin searching for?" The stories humanize Darwin and dispel the stereotype of an aged, bewhiskered Victorian. They reveal a child left to pursue his natural instincts, a buccaneering young man, a famed geologist able to explain the elevation of Patagonian terraces and the rise of the Andes.

The stories move on to tell Darwinian histories with special attention to limbs. In each one, whimsical children's literature segues into the fossil record of lobe-fins, whales, elephants, and birds—as well as the future of coral-foraging pigs. These stories traverse a landscape of Darwinian thoughts and navigate the wonderful relationships between the dead and the living. Kiplingesque questions such as "How did the whale get its mouth? The elephant its trunk?" dominate the Darwinian histories.

Shifting the Stereotype

Ask an impertinent question and you get an impertinent answer, says the noted historian of science Jacob Bronowski.[2] Darwin's imagery of the natural world often proves unsettling. For many, Darwinian histories, anchored to an unbroken chain of material causes and chance occurrences, clash with religious dogma. Where some find exhilarating insight into the complexities of the living world, others experience despondency in noting the apparent purposelessness of evolution. How impertinent, they think: Because extinct fish had two eyes, so do we as their progeny! Not to mention a brain like other primates. Darwin's disquieting thoughts and demanding, counterintuitive ideas have troubled people for generations.

To adopt Darwin's view of life is to struggle with very deep-seated, intensely human perspectives and longings that accompany the coming of age. Charles Darwin grew up in a household influenced by Jean-Jacques Rousseau and grandfather Erasmus Darwin, plus several sisters. No doubt his childhood collections and youthful circumnavigation of the globe had a great effect on his life's ambitions. As a young man "full of 'satiable curtiosity,"[3] Charles Darwin indeed asked impertinent questions—as if in search of scientifically rigorous just-so stories.

Like young Jim Hawkins, Robert Louis Stevenson's protagonist in *Treasure Island*, Darwin sailed the southern seas immersed in formidable adventures.[4] Jilted by his love, armed with a brace of pistols, sword and cutlass by his side, twenty-something Charles Darwin braved rebellion, storms at sea, and earthquake. What was he searching for amidst all of this drama?

He found true treasure everywhere, whether in the guts of Galápagos iguanas or the cell structures of encrusting bryozoans. He dug for fossils in the cliffs of Patagonia, rode with gauchos on the Argentine pampas, and marched to suppress rebellion in the cities of Spain's former colonies. He trekked through Andean mountain passes, struggled to survive the squalls and swells of Cape Horn as clouds of spray topped cliffs two hundred feet high.

Darwin knew well the maritime achievements of Captain James Cook, the first European to make contact with Australia as well as the Hawaiian Islands. He reveled in reading and rereading accounts of Alexander von Humboldt's exploits in South America, summiting peaks and ascending rivers. Sir Walter Scott's novels had piqued Darwin's romantic notions of gallantry and gentlemanly goodness in the face of strife. And Milton's poetic voice, speaking from the pages of *Paradise Lost*, fueled his desire to experience the good and evil of the world.

The *Beagle* years sculpted Darwin's identity and sense of humanity. He found unity among humans and condemned slavery. His exploits introduced him to colonial genocide and convinced him of the civilizing power of education. He bore scientific witness to the flickering blur of geologic change and delightedly dined—sometimes quite literally—on life's biodiversity. The idealistic young Darwin never refrained from condemning humanity's depravities or appreciating its civilities, derived, as he ultimately believed, from the social instincts of animals.

Had he never authored *On the Origin of Species* nor coined the phrase "natural selection," Darwin would still have secured his *Beagle* fame. Whether observing Pacific atolls, Andean ridges, or Patagonian terraces, he was, quite simply, privileged to be the first naturalist to circumnavigate the globe informed by Sir Charles Lyell's newly published *Principles of Geology*, the sourcebook for how to interpret the distant geologic past by extrapolating causes acting in the present. For a young man, promoting empire provided an opportunity to conduct cutting-edge geology and a chance to test his mettle.

As a teenager, and guided by his mentor, Robert Grant, Charles Darwin had strained tiny invertebrate life forms from the Firth of Forth in Edinburgh. Motile algae, strange tiny eggs, and tentacled polyps caught his attention. He later pursued this interest in humble marine life from ocean island to ocean island and along continental shores around the globe.

In his old age, Darwin returned to wondering about the common origin of all life, both plant and animal. A grandfather himself, he reread his own grandfather's works. Erasmus Darwin's *Zoonomia* provocatively proposed life's ascent from a primordial filament of life, a fiber neither plant nor animal but capable of becoming both.

The Origins of Origins Stories

"How did plants begin?" a child might ask. Could plants and animals have begun as one? Maybe so, speculated Robert Grant, who found in the sea sponge vestiges of life seemingly undecided about their taxonomic affiliation. Perhaps Darwin also imagined plants to be the descendants of a sea-spongish ancestor—an ancient SpongeBob SquarePants. Grandfather Erasmus Darwin thought so. Certainly, humble organisms captivated Charles Darwin's attention, especially the marine polyps of *Flustra*, a colonial animal.[5] Near the end of each of the six editions of his *Origin of Species*, Darwin mused "that probably all the organic beings which have ever lived on this earth have descended from some one primordial form, into which life was first breathed."[6]

Motility, the ability to move, impressed him. Both primitive plants and animals possessed motility and responded adaptively to their surroundings. Vine tendrils and polyp tentacles reached out in similar fashion. Had they preserved traits from life's primordial filament? Had they the rudiments of sensation and will?

Ancient creatures, he realized, resembled those in the present. Ones on distant islands resembled those inhabiting nearby land. Even though differences abounded, they seemed to follow trends in time and space. "Such facts as these could be explained on the supposition that species gradually become modified; and the subject haunted me," he confessed in later life.[7] Island-by-island variants, whether tortoises, finches, or mockingbirds, spoke to him in the same apostate voice. Mutability, they whispered forcefully, characterizes not only the rocky Lyellian earth but every living thing that dwelleth and creepeth upon it.

On board the *Beagle*, Captain Robert FitzRoy affectionately nicknamed him "Philos," short for "Philosopher." Exhumed from Philos's memories were thoughts of

> great fossil animals covered with armour like that on the existing armadillos . . . the manner in which closely allied animals replace one another in proceeding southwards over the continent . . . by the South American character of most of the productions [plants and animals] of the Galapagos archipelagos, and more especially the manner in which they differ slightly on each island of the group.[8]

Origins stories and the imagery of evolution descend from his *Beagle* voyage: the origins of creatures sporting limbs, trunks, and baleen among them, the branching imagery of common ancestry triumphant. His collections span fulgurites to finches; his island treasures enrich the world. At the center of this trove stands a very young treasure-seeking man, a compulsively inquisitive, ambitious, and idealistic adventurer grasping to understand the natural world that created him.

That's where the story of origins stories begins: stories that tell of the wonderful relationship between the dead and the living, between lobe-finned fish and humans, moose and whales, pachyderms and manatees, birds and dinosaurs.

Storytelling

According to the literacy expert Karen Gallas, children's disarmingly simple science talk raises truly fundamental questions. Whether answered in myth or science, questions about the origins of things hold deep appeal to young children. Asking "How did birds begin?" prompts big-picture thinking, while wondering "Why don't birds have teeth?" calls for explanatory reasoning.[9] Both are Darwinian questions, and speculative responses to such questions engage children in embryonic theorizing.

Perhaps the first birds were like dinosaurs. Or maybe crocodiles. (The egg did come first, by the way.) When children speculate, entertaining multiple possibilities is the goal, not quick right answers. Their storytelling resembles theory-building. It begins the search for what might count as evidence.[10] Origins questions invite exploratory thinking, and stories of origins exploit fertile imagination and metaphor. Scientific claims depend, of course, on empirical evidence processed with logic; in turn, what counts as evidence depends on a conception.[11] In the case of origins, the name for this conception is "evolution." Origins stories, in effect, provide for the reenactment of evolutionary explanations.[12]

Very many children love dinosaur stories: the stories fossils tell. What other stories are populated by monsters that even adults believe in? Wondering about extinct monsters naturally led nineteenth-century adults to questions about the history of life, and it's a good starting point for children, too. Beasts come and go, but how? They are alike and different in so many ways, and that's the first clue. Children (and adults still willing to listen to their inner child) are perfectly capable of asking charming yet philosophically and scientifically significant questions such as "Are duckbilled dinosaurs big, wingless ducks? Is a camel more closely related to a horse or to a giraffe? What makes a penguin a bird? Why is a whale not a fish? Do elephants have knees?"

Transitions by creatures from sea to land and back again capture the imagination. Rudyard Kipling's fanciful stories and their corresponding scientific questions tap the same imagination and curiosity that children possess in abundance. If Kipling were writing *Just So Stories* today, he might try to answer "How Did the Lungfish Gets Its Lungs?" Or perhaps "How Did the Tetrapod Get Its Limbs?"[13] Limbs and bodies, skin and teeth: trading the aquatic lifestyle for the terrestrial one taxes these features tremendously. Behavior and anatomy intertwine in the struggle to adapt to the changing circumstances of life.

Children's literature holds surprising potential for awakening interest in these evolutionary dramas. The true story of *Tiktaalik*, a creature known from its Devonian period fossils, is the story of how the tetrapods became limbed. The evolution from fins to limbs—the story of the invasion of land by backboned creatures—plays out implicitly in Leo Lionni's tale *Fish Is Fish*.[14]

Devonian fish skittering on stubby fins among logs in shallow water hundreds of millions of years ago bequeathed limbs to the earth's contingent of four-legged land-roaming beasts—as well as to the two-legged ones who came to stand upright. Given the unbroken descent of limbs from fins, maybe cows should be thought of as a type of lobe-finned fish with udders. Cows, on a rudimentary classification of limbed animals, do stand close to antelope, moose, and whales. Sprouting limbs opened new horizons to fish. Are people two-legged fish with feet and fingers in place of fins?

Rudyard Kipling's "How the Whale Got Its Throat" features a sagacious mariner outwitting a whale. Darwin speculated on the origins of baleen when his contemporaries objected to the viability of intermediate forms. What good is a whale's mouth not yet festooned with curtains of baleen? Darwin, in trying to imagine the origin of baleen, wondered about swimming bears as analogues for early whales. He pictured bears with their mouths agape, paddling vigorously, straining pond life like ducks. Voilà: baleen many generations later. Scaling tiny bristles in a duck's bill in keeping with the dimensions of a whale's mouth gave him that idea. In principle it works, provided that something bristle-like might be growing in bear gums. Critics forced Darwin to back away from a hypothetical bearduck as a model for the ancestry of whales in the series of editions that followed the 1859 original. Too bad. The imagery is terrific.

Paleontologists now suggest that ancestral proto-whales once walked to the ocean—perhaps like bears charging into a stream to seize migrating salmon. Where one day Pakistan's mountains would rise, wolfish predators splashed into the sea in the hunt for fish. Flash forward tens of millions of years and their legs are gone. Flukes propel them. Their backs are well-muscled, the better to let them gallop through the ocean.

Having observed their legs, hooves, and things on their heads, Bernard Wiseman's logic-obsessed moose, Morris, inferred that cows belonged, the same as himself, to the category of moose.[15] His cow companion advanced a biological argument of her own: having a cow for a mother meant she was a cow. Morris clinched his argument by applying the same logic in reverse. Since his new friend was a moose, her mother must also have been a moose.

By consistently applying logic similar to Morris's reasoning, paleontologists group whales and hippos among Morris's cow and deer friends—all fine examples of mooseness to Morris. If a whale is what its ancient mother was, does that make a whale a type of moose? This conundrum often vexes evolutionary thinking. Novelty begets newness. Relationship is the key issue: a moose is indeed related to a whale, rather surprisingly so.

Naming

Ideally, what makes a whale a whale and a moose a moose? For Herman Melville's Ishmael in *Moby-Dick*, a whale is a spouting fish with a horizontal tail. They are the "royal fish" and when secured are known as "fast fish."

Modern scientists hold a different opinion. Whales lactate—like cows—a defining attribute of mammals. For the scientist, internal anatomy reigns over the kingdom of classification. For the whaler, external appearances and behavior do (or once did). Common names such as starfish, shellfish, and jellyfish make perfectly good sense in everyday living, and whalers in Herman Melville's day clearly understood that whales were fish. On whose authority is a whale a fish? In 1818 the City Court of New York so declared, and as a result the port authorities imposed the common fish oil tax on all barrels of oil collected from whales. Playing fast and loose with the definition of a "whale"—with unsavory references to how it nursed and mated—did not impress the legal establishment.[16]

Scientific purists may recoil from this judgment and sincerely lament the labeling of whales as "fish." Such naming challenges their authority. When it comes to suggesting common ancestry, anatomical characteristics are hard to beat. When it comes to managing "fisheries," perhaps different criteria work just fine.

Categorization defines what makes a bird a bird or a dinosaur a dinosaur from a human point of view; nature is whatever nature does. The categorization schemes of evolutionary biology lead, in contrast with the everyday encounter with the world, to rather strange claims. For example: birds are avian dinosaurs. Yet, as everyone recognizes, parakeets lack much of the bravado associated with the Jurassic's *Allosaurus*.

Outsiders to any professional scientific community lack an understanding of how experts use specialized categories to achieve particular aims. As a result they find jargon distasteful and naming confusing. Without being placed in a purposeful context—and without an element of playfulness—any taxonomy appears arbitrary and arcane. Origins stories offer an escape route from this trap. They give to naming a clear purpose: inferring common ancestry.

Playful Imagery

While collecting pebbles, Leo Lionni's three young frog friends, Jessica, August, and Marilyn, find themselves surprised when a very special pebble hatches. The newborn chick immediately rushes to the pond and swims wonderfully. Its mother turns out to be an alligator. Have the little frogs stumbled onto the origin of birds as avian dinosaurs? Cold-blooded creatures from frogs to chickens lay eggs, but only some resist desiccation. Egg type seems to have split the tree of life, with froggy amphibians—and their gelatinous eggs—on one branch, and gators and birds, hatching from pebbly eggs, close to each other on another. The frog friends' perceptions resemble paleontological claims: birds and gators have much in common. They are quite different from amphibians, and they do share some egg membranes that amphibians lack.

Opportunity knocks serendipitously, as when Bill Peet's worldly pig Chester's markings open the doors to a career in the circus. Who knows what pigs of the future may look like if bred for the circus or let loose to forage among corals. If we look back from the future on the geologic timescale, will pig noses acquire the status of proto-snorkels? What might become of sheep that browse tide pools for algae? Where plastic behavior and flexible diet lead, form may follow. Chester's desire to see the world lands him in the circus, standing on his nose and displaying a map of the world on his back. If people select performing pigs, might Nature select aquatic sheep?

Viewed from just the right angle, the playful imagery of whimsical stories for children may frame the introduction of serious Darwinian thoughts. Lionni's frogs' perceptions of a scaly hatchling and Peet's pig's serendipitous circus success suggest "just how." Origins stories may extend and enrich the Kipling tradition of "just so." Good storytelling mines curiosity. It models how exuberant playfulness with ideas enriches the disciplined study of a science.

Darwinian Histories

As a form of disciplined study, Darwinian histories specify "the changes that take place from generation to generation along a lineage."[17] Darwinian histories reconstruct intermediate forms, establish their usefulness, and point in the direction of their incipient origins: the creature with an eye not yet an eye, the beast with a flexible lip not yet a trunk, the animal with a feathered limb not yet a wing.

Darwin believed that "natural selection" was his most powerful and creative insight into the descent of species. Random factors no doubt have set some pathways in motion and obstructed others. Developmental constraints and gene linkages, once in place, deny natural selection a blank slate. Each new direction is dependent upon and derived from existing forms. If it were not so, constructing Darwinian histories would be impossible.

Individual by individual, thought Darwin, selection both preserves and discards variants. Individuals vary without foresight. Even so, populations—not individuals—adapt to the changing conditions of life. This process leaves intermediate forms in its wake, their existence greatly obscured by the incompleteness of the fossil record.

Surviving is a chancy business where uncertainty reigns. All forms of life are constrained by their past yet maintain an unpredictable potential for the future. An organism gambles on survival by betting its inheritance hedged with chance mutations. Sometimes it gets lucky and squeezes through circumstances quite unlike those survived by past generations. Novelty often arises when old parts take on new uses.

Darwinian histories tell of the wonderful relationships between the dead and the living, no matter how distant. Evolution's meaning begins as Darwin phrases it in the conclusion to his final edition of *On the Origin of Species*: "descent with modification through variation and natural selection."[18] Deeply layered with surprises, his conception is onion-like: peeling away one layer only exposes another to grasp, as the inner layers of evolution ensnare who we think we are. His histories evoke tearful appreciation for the hosts of extinct beings yet promise the sweet taste of understanding life's continuity.

Children's stories offer unexpected ways to bridge the gap between fanciful imaginings and Darwin's significant ideas. Whether intentionally or unintentionally, the authors of children's books often raise questions about evolution both whimsical and serious, meriting answers both playful and scientific.

1

"CURTIOSITY'S" CHILD

Bobby Darwin's Impertinent Early Years

But there was one Elephant—a new Elephant—an Elephant's Child—who was full of 'satiable curtiosity, and that means he asked ever so many questions. And he lived in Africa, and he filled all Africa with his 'satiable curtiosities. He asked his tall aunt, the Ostrich, why her tail-feathers grew just so, and his tall aunt the Ostrich spanked him with her hard, hard claw. He asked his tall uncle, Giraffe, what made his skin spotty, and his tall uncle, the Giraffe, spanked him with his hard, hard hoof. And still he was full of 'satiable curtiosity! . . . He asked questions about everything that he saw, or heard, or felt, or smelt, or touched, and all his uncles and his aunts spanked him. And still he was full of 'satiable curtiosity!

Rudyard Kipling, "The Elephant's Child"

Ernst Mayr posed the question "What made Darwin such a great scientist and intellectual innovator?" He then answered himself, employing Kipling's famous phrase: "He was a superb observer, endowed with an insatiable curiosity. He never took anything for granted but always asked why and how."[1] And Darwin's questions were many:

> Why is the fauna of islands so different from that of the nearest mainland? How do species originate? Why are the fossils of Patagonia basically so similar to Patagonia's living biota? Why does each island in an archipelago have its own endemic species and yet they are all much more similar to each other than to related species in more distant areas?[2]

As Evolution's Child, Darwin practiced curiosity from a young age. He ventured not to the banks of the great grey-green, greasy Limpopo River, but to the shores, forests, mountains, and plains of Brazil, Patagonia, Chile, Tierra del Fuego, and the Galápagos Islands. What *was* the origin of his inquisitive "'satiable curtiosity" tendencies?

The Origins of Impertinence

Of course, no definitive answer to that question exists. Family values, childhood experience, and inherited nature interact in complex ways, placing great distance between the relatively simple just-so stories accounting for the origin of whale flukes and elephant trunks and the more complex tale of human mental development. The origin of curiosity in the mind of a human being defies complete explanation. Still, there is an abundant biographical record of Darwin's life, told by historians, his descendants (for example, Randal Keynes, a great-great-grandson of Darwin, and author of *Darwin's Daughter*), and even the man himself. His early childhood holds clues about how to promote curiosity as both a cultural value and a personal trait.

In his earliest years, young Charles Robert Darwin enjoyed a childhood of leisure and play. As poignantly emphasized by historian Janet Browne, his sisters called him "Bobby." He romped across the lands of the family's English estate, the Mount, in Shrewsbury, above the Severn River, near the Welsh border. He had one younger (Catherine) and three older sisters (Marianne, Caroline, and Susan). Caroline oversaw his education and was reportedly quite stern. Charles had one big brother, Erasmus, with whom he was very close. He often visited and played at the estate of his Wedgwood cousins, perhaps troubling his aunts, uncles, and other family members with question after question; he was the family's "Elephant's Child."

According to his biographers Adrian Desmond and James Moore, from childhood on Charles "craved" praise and acceptance. "He was an inveterate collector and hoarder—shells, postal franks, birds' eggs, and minerals. They were trophies, piled up for praise."[3] He feared provoking displeasure, yet often could not resist the temptation to engage in mischief, then tried to cover up his backyard antics.

Young Bobby roamed, collected, and spent many afternoons fishing with worms. Unlike the marine invertebrates that would captivate him later on, his terrestrial worms lacked the ability to desalinate water. He quickly learned to euthanize them in a saltwater bath. Late in life he would author his final scholarly tome, *The Formation of Vegetable Mould, through the Action of Worms, with Observations of Their Habits* (1881): a worm scientist when just seven as well as when seventy.

At home he marveled at the beauty of and differences among the pigeons kept by his mother, Susannah Wedgwood Darwin. His appreciation of pigeon fanciers and their broods would yield the most fundamental insight into his solution to the problem of the origin of species: the analogy between variation under domestication—scaly and feathered pigeon legs, for instance—and variation within natural populations of organisms, such as stocky and slender finch beaks. His preschool collecting days disposed his mind to attend to nature's endless capacity for producing variation. His childhood encouraged him to experience delight and joy in such details. Death, however, soon intruded upon his idyllic existence.

The shocking loss of his mother in the midsummer of 1817 came when Charles was only eight and had just begun formal schooling. His older sisters had to assume even greater responsibility for his care. The following year found him bound with affection and admiration to his thirteen-year-old brother, Erasmus ("Ras," who later became a favorite uncle to his children), a fellow boarder at the Shrewsbury School. By the time he was ten, rote learning, recitation, and authoring verse dominated school life. Yet Charles found the opportunity to read great literature—Byron, Shakespeare, and Horace—with pleasure. He continued to amass collections and take long, solitary walks, "day-dreaming of tropical islands and South American landscapes . . . a refuge from the stultifying grown-ups."[4]

Upper-class education for English youth in the nineteenth century did not include the study of the sciences. The Shrewsbury boarding school, which he attended from 1817 to 1825, stressed drill and the classics. Teenage Charles found refuge from tedium in the family estate's garden tool shed. Here, with his brother Ras, Charles pursued an avid interest in "chemistry." They assembled the requisite retorts, beakers, and mortars and pestles, obtained a great variety of reactive substances, and actively conducted chemical experiments in their makeshift lab—not always with complete safety in mind. The boys combined their reagents to make new substances: solid, gaseous, and liquid.

Some reactions required heating, and sometimes they supplied more than necessary. As a result, he and Ras frequently ignited things, not just in their backyard tool shed but also at night in the boarding school dorm, where he experimented

Figure 1.1 Young Charles Darwin destined to become a worm scientist. Illustration by Jan Glenn.

with an open flame. Some of the vapors created by his chemistry set ignited quite well, and teasing soon earned schoolboy Bobby the nickname "Gas."[5]

Chemical experimentation may have seemed a relatively novel hobby, but experimentation with kiln-firing techniques was actually a natural extension of family tradition. Darwin's maternal grandfather, Josiah Wedgwood, built his fortune on industrial-scale pottery-making, and his innovative manufacturing company rose with the tide of England's Industrial Age.[6]

Beyond material riches, the Darwin family was heir to the ideas of the Enlightenment, especially the thinking of Jean-Jacques Rousseau, and they trusted in the investment of capital to return increasing wealth. The Darwins and Wedgwoods were, in this way, classic liberals. They believed in upward progress through competition and industrial innovation—a global process properly led, to their way of thinking, by England. Charles Darwin was a third-generation freethinker, responsive to egalitarian ideas despite being a member of privileged society.

Rousseau's Darwin

Devotion to the virtues of learning spanned the generations of the Darwin family. Darwinian upbringing stressed childhood exploration and interest in the natural world, family values Charles would refine into meticulous investigation and provocative theorizing. Grandfather and physician Erasmus Darwin, who numbered among his close friends the steam engine inventor James Watt and the chemist Joseph Priestley, spent a lifetime engaged in scientific debate. He held no small interest in the origins of life's novelties, and his influence on his son Robert Darwin made medicine a family profession, a tradition that was broken when Charles Darwin enrolled in medical school but reacted with revulsion to human surgical dismemberment.

Dissenter from the Anglican faith, freethinker, Unitarian, and French egalitarianism sympathizer, Erasmus Darwin had worked out his own ideas about inheritance and the transformation of life in his *Zoonomia*. In 1837 Charles Darwin made reference to this title in his private notebook, where, in conversation with himself (to avoid the risk of censure and disapproval), he had begun to elaborate upon his thoughts about the "transmutation of species," or one kind of creature becoming something entirely different.[7]

Darwin family tradition reinforced the value of playful, even risky exploration in thought as well as empirical experimentation with things. Charles's father, an intimidating man of massive physique, may have scolded his son from time to time on account of his backyard ramblings and incessant collecting, but both the "Darwins and Wedgwoods had a long-standing interest in advanced approaches" to rearing children, and many of these stemmed from the teachings of Rousseau.[8]

Ideas from Rousseau's *Émile* clearly influenced Erasmus Darwin regarding the education of children. He promoted Rousseau's views when he advised his own daughters in their planning of a boarding school for girls. Grandson Charles, who honored his grandfather with a biography, acknowledged his debt to the elder man's thoughts about human nature. Quoting Erasmus, he noted, "A sympathy with the pains and pleasures of others is the foundation of all our social virtues."[9] To Charles Darwin, this statement was profound. It implied that morality and ethics did not originate through divine or supernatural intervention. Instead, they resulted from sympathy for and empathy with others. But where might such sympathy and empathy have come from?

At the peak of his intellectual journey, Charles Darwin concluded that this same sympathy and empathy with others was no more than the human expression of social instincts observed in non-human animal behaviors. He believed that these instincts were inheritable and evolved under the influence of natural selection working to promote group survival through cooperation. This line of reasoning reconciled the strife, ugliness, and pain of life with its equally real exquisite, sublime, and noble experiences. In time, and by virtue of natural selection, *selfishness begot selflessness*. Selection converted the amoral struggle to survive into social instincts; human learning then expanded on these to achieve social virtues.

Across the generations, recognition of the parallelism between social virtues and social instincts united the minds of grandfather and grandson. This view provided a cornerstone to Darwin's thinking about the origin of civilized behavior and conflicted with contemporary theological accounts of virtue. The England of Darwin's youth was not a modern secular state. Anglicanism reigned, and the joint role of church and state was accepted as indispensable to the establishment of a just and moral society.

Rousseau's and Darwin's views accelerated the transition from a sectarian to a secular society. These views discounted divine intervention, whether as creation, grace, or soul, as the source of civilizing virtue. There was no room for original sin. Rousseau taught unequivocally that people are born good, free of sin: "Everything is good as it leaves the hands of the author of things; everything degenerates in the hands of man."[10] Therefore there was no need for harsh intervention in the Darwinian view of early childhood education. Instead there was a call for learning to sympathize with the thoughts and feelings of other humans and the rhythms and details of the natural world—free from adult imposition.

Rousseau's philosophy of childhood education also influenced the Wedgwood household, where young Emma, Darwin's first cousin and future wife, along with her brothers and sisters,

> were "allowed to act in an unrestrained manner without rules and precepts," as their father felt that "every act of interference does harm" to a child's

> nature. "The children may be taught to exercise their faculties by inducing them to answer their own questions, either experimentally, or by having the subject so presented to them that the inference shall be sufficiently clear without its being drawn for them."[11]

Johann Heinrich Pestalozzi championed Rousseau's view of childhood and promulgated methods of instruction that encouraged learning from direct experience.[12] His thought was current in the early 1800s, and Darwin's sisters Caroline and Susan, following the death of their mother, turned to Pestalozzi for guidance in raising their younger siblings.[13] Pestalozzi concluded that "the child must be led to see for himself that which he is to learn, and not to take it upon the mere authority of the teacher."[14]

According to Darwin's biographer and descendant Randal Keynes, "Charles and Emma's approach with their children was undemanding and liberal; they saw little value in discipline and learning by rote, but wanted to encourage their children to think for themselves."[15] Charles Darwin's grandson Leonard, whose mother died from childbirth complications and who came to live in his famous grandfather's household, assimilated this undemanding and liberal approach quite well. As Irving Stone recounts in his "biographic novel" about Darwin:

> Once when he wandered into the sitting room Charles found five-year-old Leonard jumping up and down on their red velvet sofa.
>
> "Leonard, I've told you I didn't want to see you jumping on that sofa."
>
> "Well, Papa, if you don't want to see me, I suppose you'll just have to leave the room."[16]

A habit of collecting also stayed at the heart of Darwin family early childhood experience through the generations. Darwin's children learned to collect for themselves, finding special joy, and a fitting mode of rebellion, in capturing a species unknown to their famous beetling father: His [Charles's] boys were butterfly hunters. Like most young and ardent lepidopterists, they despised the beetle collectors. It was also a way of pulling their father's leg. By precept and enthusiasm he had instilled in his children a taste for natural history, but he knew better than to give them specimens. Youngsters had to collect for themselves to acquire the fervor.[17]

Nature and Nurture

To be human is to be potentially curious, even impertinently curious—to question and take little for granted as a child. Yet this style of behavior depends on a cultural context enacted through the family. Cultural context, if it is to yield a Darwinian propensity for asking how-and-why questions, must nurture not only the questioning but also the freethinking that permeated the Darwin and Wedgwood households generation after generation. While the potential for being curious may reside deep within human instinct, insatiable curiosity is a cultural value, not something automatic. We may trust in the equal potential of all children to develop their curiosity without overly romanticizing early childhood and fully accepting Rousseau's philosophy of education. Darwin children were not left alone; they were immersed in experiences, interests, language, and examples valued by their family.

Emma and Charles Darwin accepted the German Romantic novelist Jean Paul Friedrich Richter's dictum, following Rousseau, that "play is the first poetry of the human being"[18] Charles Darwin never departed the playground of science. From childhood beetle collections through garden shed chemistry; to *Beagle* adventuring around the globe in search of fossils and finches, tortoises and iguanas; to barnacle and worm dissection; to pigeon breeding and animal husbandry; to orchid hybridizing and vine cultivation; to decades of adult journaling, science society debating, and international publishing; to accounting for the descent of man and theorizing (incorrectly) on mechanisms of inheritance, he innovatively manipulated ideas and things. His youthful explorative playfulness matured into disciplined thought, ever anchored in impertinent questions. He was an "Elephant's Child" until his death.

In the wake of his childhood opportunities shaped by Rousseau's educational precepts, Darwin certainly received formal training and pursued rigorous traditional academic studies, beginning with study of the classics at the boarding school for young gentry in Shrewsbury. Great scholars, teachers, and scientists took him under their wings. First, he followed in the footsteps of grandfather Erasmus, father Robert, and brother Erasmus to Edinburgh University, with the intent to become a third-generation medical doctor. Though disenchanted by human dissection, he did find stimulating the work of Robert Edmond Grant, who adhered to the transmutationist ideas of Jean-Baptiste Lamarck. Both were enamored with the biology of sponges and had concluded that natural law, played out through climate and environment, drove a process of evolution from simple to more complex forms.[19]

Later he attended Christ's College, Cambridge. Before departing on the voyage of the *Beagle*, he learned taxidermy from a freed Afro-Caribbean slave and taxonomy from a variety of Cambridge dons. While in port in South America, he had delivered to him the latest edition of Charles Lyell's *Principles of Geology.* (The captain of the *Beagle*, Robert FitzRoy, had provided him with a gift of the first edition upon their departure.)[20]

Throughout his life he maintained a copious correspondence with leading experts in many fields. His estate in the English countryside was a virtual biological research station: pigeons, vines, worms, barnacles, germinating seeds and orchids all under scrutiny. Darwin was no stranger to formal knowledge and disciplined thought, but his intellectual adventure began with playful exploration and passionate collecting.

Impertinent Humanist

Though he wished to evade spankings, literal and figurative, Darwin grew to ask the most impertinent question of all, "How did we become human?" His answer—through descent with modification by means of natural selection from ancestors we hold in common with other species—reduces to a single word: evolution. Overstating the importance of this concept is difficult. "Evolution is the most profound and powerful idea to have been conceived in the last two centuries," wrote Jared Diamond in 2001.[21] Whether humans engineer new genomes at the tiniest scale of life or alter habitats on the scale of continents, incumbent upon us is an understanding of how evolution, with and without human direction, works. What transpired to fashion a wealthy young boy's curiosity and pleasure into such an important idea? Apparently the process began with ample opportunity for playful exploration. He sought praise and acceptance, but not at the expense of abandoning his passion for questions. His desire for praise only made him work harder to get his answers right.

Evolution happens, according to secular and materialist theory, toward no end, for no purpose, and without meaning. Such a story offers little comfort. Evolution reveals life's connectivity and creativity and has great potential to induce awe. Evolutionary sciences, however, offer no recipes for obtaining meaning from life or for finding ultimate understandings of purpose.

The Darwin household encouraged freethinking; Darwins and Wedgwoods challenged dogmas. Charles Darwin might likely concur with the evolutionist Richard Dawkins's provocative pronouncement "Religion teaches you to be satisfied with wrong answers. It's sort of a crime against childhood."[22] Yet the religious roots of the Darwin and Wedgwood households in the Unitarian and Anglican churches reinforced his cherished belief that all humanity constituted a single species, the distances dividing peoples stemming principally from upbringing and education.

Many find Dawkins's atheism rather offensive. In milder terms, the education of Charles Darwin demonstrates the value of impertinence—of encouraging children to ask, as did the Elephant's Child, impertinent questions. "That is the essence of science: ask an impertinent question, and you are on the way to the pertinent answer."[23]

Darwin was well on his way to a pertinent answer to the mystery of mysteries. For him, the histories of species were not merely just-so stories. They constituted explanations cast in natural causes that account for life's diversity within its unity.

Life, in all its complexity, did spring from primitive precursors. Among ancient populations of lobe-fins, skulls grew flat. Their descendants developed thickened necks; brains became bigger and limbs stronger. Having descended from fish—and from something resembling modern fish-like, filter-feeding, segmented, back-stiffened, blade-shaped, undulating marine lancelets before that—these creatures became hippos and moose and whales and elephants and people, however impertinent this truth may seem.[24]

Darwin's privileged life provided him opportunity and his Victorian age a social opening for this conclusion. His family culture valued an approach to learning that would profit his intellectual journey immensely. To the education of the modern evolutionist, the study of Darwin's own life and words remains pertinent indeed. They defang the image of mythical monster, temper the portrait of solitary genius, and reveal a person of human scale devoted to the passionate pursuit of interest.

2

DARWIN AND THE PAMPAS PIRATES

Adventure in Search of Treasure

> *His stories were what frightened people worst of all. Dreadful stories they were—about hanging, and walking the plank, and storms at sea, and the Dry Tortugas, and wild deeds and places on the Spanish Main. By his own account he must have lived his life among some of the wickedest men that God ever allowed upon the sea, and the language in which he told these stories shocked our plain country people almost as much as the crimes that he described.*
>
> Robert Louis Stevenson, *Treasure Island*

Thus young Jim Hawkins describes "the captain," the mysterious guest lodged at the Admiral Benbow Inn, whose death sets in motion the events in the adventure classic *Treasure Island*. Each evening as the captain, soon revealed to be the pirate Billy Bones, descends into a drunken stupor, he presciently chants, "Fifteen men on the dead man's chest—Yo-ho-ho and a bottle of rum!" Billy Bones has with him a chest, and in the chest lies a map of hidden treasure marked with an "X." As Jim soon learns, many nefarious characters covet the map, among them "a seafaring man with one leg."[1] Too much rum and excessive fear doom Billy Bones. He dies of a stroke, and the chest becomes the property of the inn's proprietor, Jim's mother.

The map holds the key to untold riches—a bounty accumulated and secreted away by the deceased pirate Captain Flint. (Pirate captains do not fare well in Stevenson's tale.) Flint's men mean to recover the treasure. Black Dog has sniffed its trail, and Billy Bones has warned Jim to be on the lookout as well for the menacing Long John Silver—the seafaring man with one leg.

Pirates always seem to lack body parts—often an eye or a leg—and this duo is no exception. Two fingers are missing from one of Black Dog's hands. Long John's leg ends just below the knee, causing him to hobble about using a crutch. A parrot always perches on his shoulder. The image of Silver terrifies young Jim Hawkins.

It is Silver who looms largest in this tale of daring, betrayal, and redemption. An ambiguous villain, he is both a scoundrel and the man responsible for saving the protagonist, Jim. Silver worms his way into the graces of Squire Trelawney and Dr. Livesey, the "good guys" who secure the *Hispaniola* for a voyage to recover Flint's hidden treasure. They invite Jim aboard as the cabin boy. For Jim, a coming-of-age adventure begins to unfold. Long John Silver, mistakenly trusted by the squire and the doctor, earns a place on the *Hispaniola* as cook. He graciously secures many other able-bodied seamen for the crew—in truth, remnants of Captain Flint's gang of cutthroats.

Over a century after *Treasure Island* was first published, popular culture continues to reinvent tales of treasure maps and not-entirely-villainous pirates. In Steven Spielberg's film *The Goonies*, for example, several teenagers from Astoria, Oregon, discover a map that leads to One-Eyed Willie's treasure.[2] One of the fugitive Fratelli brothers, the muscular "Sloth" of fearful physiognomy and missing teeth (he speaks somewhat poorly as well), rescues the gang and belies his supposed role as a scoundrel. Similarly, in Disney's *Pirates of the Caribbean* franchise, Captain Jack Sparrow exploits ambiguity in the dual role of pirate and hero.[3] *Treasure Island* solidifies virtually every stereotype of a pirate adventure imaginable. Ironically, the story has the power to undo the stereotypical image of Darwin as an aging, dyspeptic sage. There's quite a bit of Jim Hawkins in his youthful persona, and perhaps a bit of romantic admiration for the piratical characters of the world.

Pirate Darwin?

Villains embarking on high seas adventures, gentle-natured giant scoundrels, and ambiguously heroic pirate captains seldom serve as an introduction to Charles Darwin. Yet it's time they did. When imagining Charles Darwin, instead of a Victorian gentleman, think young Jim Hawkins. Dreams of finding adventure and fame in tropical latitudes, tracing mysterious Patagonian rivers to Andean sources, and rounding the Horn into the maw of the brutal Southern Ocean no doubt animated the young Charles Darwin in much the same fashion that Black Dog's map fascinates young Jim Hawkins.

Picture evolution's swashbuckling hero on the deck of the *Beagle* amidst a stormy sea. The eerie bluish-green light of St. Elmo's fire illuminates the mast. Waves crest and splash over the deck. A pistol and cutlass hang from his belt. Darwin at that moment is eager to join a shore party in pursuit of rebellious locals. He gazes through his spyglass at an albatross maneuvering in the rising gale. A sassy bird perches on his shoulder—perhaps a Galápagos finch—or, better yet, a mockingbird. At least not a parrot, Long John Silver's choice, forever chanting "Pieces of eight, pieces of eight" in mocking memory of Captain Flint. One can almost hear Darwin's bird similarly chanting "Common ancestor, evermore!" to the consternation of Darwin's own anti-evolutionist Captain FitzRoy, a man who lived to regret escorting Darwin to the treasured islands of the Galápagos.

Figure 2.1 Pirate Darwin with cutlass and spyglass. Quoth the mockingbird, "Common ancestor, evermore!" Illustration by Jan Glenn.

Charles Darwin leaves Plymouth harbor aboard the *Beagle* on December 27, 1831, the *Beagle's* second voyage, for five years of seeking nature's secrets, five years of facing storms at sea and facing down rebels on land—five years to pirate away treasure. Such treasure! Octopuses shifting colors, puffer fish squirting water, and megabeast fossils emerging from cliffs. He finds clues to the mystery of life's origins in the branching forms of tiny coralline sea creatures. His collections embody nature's exuberance past and present—earth's ever-evolving bounty. From his voyage descend stories that still shock plain people, new ideas feared both then and now.

Good pirate tales temper lead characters' heroism and villainy with ambiguity. Did Long John Silver's heart beat sincerely as he rescued young Jim Hawkins, or was the old pirate cynically doing what he must to survive? History paints an equally mixed portrait of the progress of evolutionary thinking in its crisscrossing of the realms of organismal biology and human society. Although his exploits will yield insights into the grandeur of life, the imagery of "survival of the fittest" will also spawn the pernicious dogma of social Darwinism, a source of barbarism no better than piracy.

Courting Fuegians

Outfitted for making maps and charts, the *Beagle* sets out on May 22, 1826, on a maiden voyage that will last four years. In 1828 Robert FitzRoy, age twenty-two, replaces the ship's captain, Pringle Stokes, who has gone mad and shot himself. Charting the channels of South America's extremities for the British Admiralty proceeds for two more years. Apparently, it can be depressing work. On this first voyage the *Beagle* cruises among dank islands while smaller craft busily maneuver close to shore. The men mark shoals and depths with precision among the passages surrounding Tierra del Fuego and Cape Horn. The whaleboat crews do the close-to-shore work essential to nautical mapmaking.

It is not their surveying mission, however, but an act of piracy—hostage-taking—that sets the stage for what will become the *Beagle*'s second voyage with Charles Darwin on board.

Awaking onshore one morning to find their whaleboat missing, a crew assembles a serviceable craft from branches and tent canvas. They manage to paddle back to the *Beagle* and report the presumed theft.

The loss of even one whaleboat is intolerable and jeopardizes the mission. The theft incenses Captain FitzRoy. He determines to pursue local inhabitants thought to be guilty of the crime. In short order the Englishmen storm a campsite littered with what appear to be artifacts from the whaleboat. Powerfully built Fuegians resist, hurling stones with deadly effect. They grab one crew member and attempt to smash his skull; a gunshot saves the mariner yet regrettably kills one of the suspected thieves.

Subsequently, Captain FitzRoy takes several Fuegians hostage. He intends to use them as bargaining chips for his stolen whaleboat. At one point he holds eleven captives, including three women and six children.[4]

Some of FitzRoy's prisoners attempt to describe where to look for the stolen whaleboat, but language proves an insurmountable barrier, and the boat remains missing. FitzRoy's week of intense searching through a "labyrinth of coves and channels" proves fruitless.[5] Meanwhile, except for three of the children, his captives escape. Having given up the search, FitzRoy decides to keep one child as a hostage and return the other two to their people.

The craft used by the stolen boat's crew to save themselves is little more than a seaworthy basket. In its honor the hostage child becomes "Fuegia Basket." Yok'cushly—an approximation of her given name—is about nine years old, according to Captain FitzRoy.

Days later, having commandeered a native canoe, FitzRoy persuades a twenty-something Fuegian man, El'leparu, to board the *Beagle*. Perhaps El'leparu will become an interpreter and negotiate the return of the still missing whaleboat. Nearby looms the rock outcrop the men call York Minster for its resemblance to its namesake cathedral in England. FitzRoy dubs his ersatz interpreter "York Minster."

Nearly a week later, smoke draws FitzRoy's attention to a nearby cove. An approach to the encamped Fuegians again turns violent, and a well-aimed stone strikes a crew member's head. Canoes at the camp appear to hold gear from the missing whaleboat, thinks FitzRoy. Continued pursuit leads to the capture of another Fuegian, actual name unknown, now christened "Boat Memory," for memory was all that was left of the whaleboat.

More than a month passes. As the *Beagle* nears the Murray Narrows, Fuegians approach in three canoes, eager to exchange goods. FitzRoy trades a mother-of-pearl button for O'run-del'lico, a boy of fourteen. The crew quickly gives him the name "Jemmy Button." The complement of human cargo complete, the captain hatches a plan to deliver Jemmy Button, York Minster, Fuegia Basket, and Boat Memory to the British court, educate them at his own expense, and then return them to Tierra del Fuego. There, as English-speaking middlemen (and woman), they will, given good fortune, promote his nation's trade and military interests.[6]

FitzRoy's belief in progress blinds him to the immorality of his kidnappings. To his mind, progress justifies the use of force to instill English ways among his captives. By virtue of the acquired trait of Englishness, he believes, the temporary captives may return to their homes as agents of empire able to enjoy the profits of trade.[7] As an added benefit, they might secure their own salvation by becoming members of the Church of England.

FitzRoy extolls his plan to his commanding officer, writing in a letter to Captain Phillip Parker King on September 12, 1830:

> Sir, I have the honour of reporting to you that there are now on board of his Majesty's sloop, under my command, four natives of Tierra del Fuego. . . .
>
> I have maintained them entirely at my own expense, and hold myself responsible for their comfort while away from, and for their safe return to their own country: and I have now to request that, as senior officer of the Expedition, you will consider of the possibility of some public advantage being derived from this circumstance; and of the propriety of offering them, with that view, to his Majesty's Government.
>
> I am now to account for my having these Fuegians on board, and to explain my future views with respect to them.
>
> In February last . . . I sent Mr. Matthew Murray (master), with six men, in a whale-boat, to Cape Desolation; the projecting part of a small, but high and rugged island, detached from the main land. . . .
>
> Mr. Murray reached the place, and secured his party and the boat in a cove near the cape: but during a very dark night, some Fuegians . . . approached with the dexterous cunning peculiar to savages and stole the boat. . . .
>
> Mr. Murray and his party formed a sort of canoe, or rather basket, with the branches of trees and part of their canvas tent, and in this machine three men made their way back to the Beagle. . . . A chase for our lost boat was begun, which lasted many days, but was unsuccessful in its object, although much of the lost boat's gear was found, and the women and children of the families from whom it was recovered, were brought on board as hostages. . . .

> Our prisoners had escaped, except three little girls, two of whom we restored to their own tribe, near "Whale-boat Sound," and the other is now on board.
>
> From the first canoe seen in Christmas Sound, one man was taken as a hostage for the recovery of our boat, and to become an interpreter and guide. He came to us with little reluctance, and appeared unconcerned.
>
> A few days afterwards, traces of our boat were found at some wigwams on an island in Christmas Sound, and from the families inhabiting those wigwams I took another young man, for the same purpose as that above-mentioned. . . .
>
> Some time afterwards, accidentally meeting three canoes, when away in my boat exploring the Beagle Channel, I prevailed on their occupants to put one of the party, a stout boy, into my boat, and in return I gave them beads, buttons, and other trifles. Whether they intended that he should remain with us permanently, I do not know; but they seemed contented with the singular bargain. . . .
>
> When about to depart from the Fuegian coast, I decided to keep these four natives on board, for they appeared to be quite cheerful and contented with their situation; and I thought that many good effects might be the consequence of their living a short time in England. . . . They understand why they were taken, and look forward with pleasure to seeing our country, as well as to returning to their own.
>
> Should not his Majesty's Government direct otherwise, I shall procure for these people a suitable education, and, after two or three years, shall send or take them back to their country, with as large a stock as I can collect of those articles most useful to them, and most likely to improve the condition of their countrymen, who are now scarcely superior to the brute creation.
>
> I have, &c.
>
> ROBERT FITZ-ROY[8]

Philos

Thus dedicated to repatriating "his" Fuegians, as well as to establishing a mission in their homeland, FitzRoy plans a second *Beagle* voyage. It is to this voyage of repatriation that Darwin receives his life-changing invitation. Following an interlude of English education, including an audience with Queen Adelaide, FitzRoy's Fuegians—chaperoned by a missionary minister and joined by our young gentleman naturalist—depart Plymouth as the year 1831 comes to a close. They are going home.

FitzRoy's scientist companion at once turns his attention to the natural phenomena that abound around him. He even trolls for plankton behind the ship. Captain FitzRoy addresses him affectionately, yet with a bit of a teasing tone, as "my dear Philos," an abbreviated form of "philosopher"—a rather non-pragmatic occupation aboard a survey vessel.

Darwin is young, a bit naïve, often brash, and seldom fearful. The sailors on the *Beagle* find him a bit odd, what with all the rubbish he brings on board. Writing in a youthful voice both sociable and eager to please, Darwin acknowledges the strangeness of his occupation as a "naturalista" not only to the crew but also among local populations.

The stereotype of a bewhiskered, intellectually tormented Darwin ought not to eclipse the *Treasure Island* flavor of his extraordinary, life-threatening escapades, of how his boyhood dreams became a maturing man's delight—a delight anchored in the wild taste of physical and intellectual freedom: the story of a Darwin on horseback accompanied by fierce gauchos, a Darwin mustered to march shoulder to shoulder with British marines, a Darwin crewing a tall ship through stormy waters, a Darwin taking pleasure in sleeping on the open pampas and feasting on puma, a Darwin camped near a calving glacier, watching whales spout.

What does Philos find? Luxuriant plants festooning oyster shell terraces. Petite "ostriches" prowling arid lands. Fossilized bones protruding from shelly cliffs. Brazilians, Fuegians, Patagonians, gauchos, Tahitians, Maoris, Aborigines, Africans, and Europeans mingling treacherously. Rheas running and tuco-tucos burrowing. Williwaw winds blasting over seas glowing in bioluminescence. A city crumbling from earthquake; a tsunami devastating a coastal town. Volcanic islands sinking in slow motion; Osorno's fountain of fiery rock lighting the night sky.

In search of self, Philos spends a half decade in wide-eyed adventuring. He girdles the globe with a modern scientific library in tow, a fine microscope in hand, all the while talking science, gentleman to gentleman, providing grist for his mind's mill. Despite the tribulations of disease-carrying bugs biting in the night and the need for elixirs of port and cinnamon to cure inevitable sicknesses of the bowels, young Jim Hawkins would have envied him.

No longer is he a gentleman hunter out with his cousins for partridge and snipe, or a somewhat disappointing second son in danger of wasting the family fortune on frivolous pursuits. In South America and across the Pacific, travel unleashes the inveterate childhood collector, plant lover, and hoarder of shells, birds' eggs, and minerals to pursue bigger game He's gathering stories that frighten plain folk: stories of lost love, frequent rebellions, brutal living conditions, gaucho militias, and buried beasts. Charles Darwin is on the trail of monsters from evolution's id, riding among some of the wickedest and cruelest men ever known. But monsters, at least for the moment, can wait.

Upon reaching Brazil, Darwin sets his mind on the romantic attractions of tropical climes and the promises of seafaring with a noble captain. Yet he entertains thoughts of romance of a more literal sort as well. He holds dear the memory of a day in autumn a few years earlier, of galloping into the forest ostensibly "to hunt" with his beloved Fanny Owen. Darwin recalls steadying her as she pulls the trigger and the rifle recoils, bruising her shoulder—clearly visible when bared. His biographers comment, "Teased and tantalized by this raven-haired beauty, [he] was infatuated."[9]

Memories of the fair maiden, presumably dutifully awaiting him at home in England, mesmerize him as he arrives in South America and exults in Brazil's lush natural world. The tropical vegetation of the Mata Atlantica inspires in Darwin a "rapture of raptures." He writes:

> The mind is a chaos of delight, out of which a world of future & more quiet pleasure will arise. . . . [W]andering by myself in a Brazilian forest . . . it is hard to say what set of objects is most striking; the general luxuriance of the vegetation bears the victory, the elegance of the grasses, the novelty of the parasitical plants, the beauty of the flowers. . . .—A most paradoxical mixture of sound & silence pervades the shady parts of the wood.[10]

Within a few days Darwin finds it difficult to keep his dignity while walking the streets of Bahía, for this is the season of Carnival. "Dangers consist of being unmercifully pelted by wax balls full of water & being wet through by large tin squirts."[11]

Upon returning to the "luxuriant" Brazilian forest, he becomes fascinated by ubiquitous ants: "On first entering a tropical forest one of the most striking things is the incessant labor of ants. The paths in every direction are traversed by hosts of them carrying parts of leaves larger than themselves & reminding one of the moving forest of Birnam in Macbeth."[12]

Leaf-cutter ants—a single colony of up to 8 million individuals—carry bits of leaves to nourish their subterranean gardens of fungus. The bits add up. Ant foraging equals the combined effects of all the vertebrates feeding in the forest.[13] Ants do move the forest—from aboveground to their buried chambers.

Even things much tinier than ants grow well in Bahían forests: infectious organisms. A thorn wound to the knee leads to infection for Darwin. Bedridden and very ill, he examines another of his many treasures, the puffer fish, *Diodon*, with its upside-down swimming movements and its unexpected means of defense: "It can bite hard & can squirt water to some distance from its Mouth, making at the same time a curious noise with its jaws."[14]

Most notably, Darwin, unable to exercise proper deference, manages to offend Captain FitzRoy during a discussion of slavery. Darwin's defiant, cocky attitude almost brings his participation in the voyage to an end. He sneers at the testimony of slaves to the effect that they prefer servitude to their freedom to return to Africa. Such statements are not to be trusted when offered in the presence of their master, claims Darwin. Captain FitzRoy finds being contradicted to constitute insubordination. Nearly banished from the voyage for his offending comments, Darwin tenders his apologies. (FitzRoy ultimately turns anti-slaver.)

The *Beagle* departs Bahía for Rio de Janeiro, sails past the shoals of the Abrolhos Islands, where Darwin remarks that "almost every stone has an accompanying lizard."[15] Spiders and rats are equally abundant. Darwin must perform a daring feat as the *Beagle* enters Rio's harbor. Captain FitzRoy commands all available hands to look smart as he strives to impress the men of the British fleet already at anchor. Darwin stands with "a main royal sheet in each hand and a top mast studding sail tack in his teeth" as the *Beagle* comes about.[16]

Maneuvers complete, the commander of the British fleet informs Captain FitzRoy of political unrest ashore. Defensive positions are taken by the ship's crew and marines are readied for a landing to quell a local mutiny. "I snatched a cutlass from the pile, and someone, at the same time snatching another, gave me a cut across the knuckles which I hardly felt. I dashed out of the door into the clear sunlight. Someone was close behind, I knew not whom," said sailor Jim Hawkins at a similar moment.[17] Darwin steels himself for action; fortunately, none unfolds.

Instead, he takes up residence near Rio with ship's artist Augustus Earle, the young girl Fuegia Basket, a sergeant of the ship's marines, and his good friend from the crew, Philip King.

Rio is a mail stop. In her letter to him on his departure, Fanny had written flirtatiously, sharing her hope "that there was not to be an end" of their times together as "Housemaid and Postillion."[18] In Rio, he imagines, letters expressed in her coquettish hand await him.

Rio proves to be paradise found but love lost. On April, 5, 1832, Darwin opens a letter from his sister Catherine informing him of the betrothal of his heartthrob to "a rich varmint-like character," Mr. Biddulph.[19] The indestructible Darwin weeps with vulnerability as he writes, "If Fanny was not perhaps at this time Mrs. Biddulph I would say poor dear Fanny till I fell to sleep."[20] The tortuous course of their romance, and her serial flirtations, has come to an end. Some days later he ascends the two-thousand-foot high "Caucovado" (Corcovado) and comments, "In few places could a more horrible lovers leap be found."[21]

Forgoing a lover's leap, yet hoping to erase the pain of rejection, Darwin heads into the interior with a set of disreputable companions and experiences the horror of the mistreatment of slaves. Joining the Irish settler Patrick Lennon and a despicable Scottish slave merchant, an entirely "*selfish, unprincipled* man," he treks northeast of Rio to Lennon's *fazenda* on the Rio Macaé at Socégo.[22] Roadside crosses mark where human blood has been spilled. Darwin is aghast when, during a dispute with his plantation overseer, Lennon threatens to separate all the women and children slaves on his estate from their husbands and fathers in order to sell them at auction.

The love-forsaken Darwin attempts to drown his sorrows with his travels, only to experience once again the "horrors of illness in a foreign country" as well as "the curative powers of cinnamon and port wine."[23] All such deprivation and discomfort prove well worthwhile, and he admits to "an utter loss as to how sufficiently to admire . . . the entangled mass of a tropical forest with its infinite numbers of lianas & parasitical plants & the contrast of the flourishing trees with the dead & rotten trunks."[24] He finds himself "frequently in the position of the ass between two bundles of hay—so many animals do I bring home with me."[25]

Disaster nearly strikes soon after his return from his forest adventures: large waves almost swamp Darwin's whaleboat during a landing in Botafago Bay. Soon afterwards, three members of the *Beagle*'s crew succumb to malaria; among the dead is the popular thirteen-year-old Midshipman Musters. Two months of exploring and collecting follow. Disease and rebellion, slavery and cruelty, have become daily rhythms. Even onboard ship, natural history dramas play out. Noting that the "rigging was coated with Gossamer web" deposited by innumerable spiders, Darwin carefully watches as wasps sting their spider prey.[26]

Darwin finds lots of spiders but has lost his Fanny. The Atlantic crossing and introduction to Brazil, to its rapturous forest and cruel society, come to a close.

Blancos and Colorados

As the *Beagle* departs Rio for Montevideo, Uruguay, penguins make wakes through bioluminescent algae illuminating the sea and "the night presents a most extraordinary spectacle." In the darkened sky, streaks of "the most vivid lightning" explode in the night. These "natural fireworks" welcome the *Beagle* to the Rio de la Plata. Her masts shimmer in the phosphorescent green of St. Elmo's fire. Amidst "heavy squalls of rain & wind" the *Beagle* drops anchor.[27]

As in Rio de Janeiro, threats of local political violence greet the *Beagle*'s arrival. The "Banda Oriental" (Darwin's name for the region that would become Uruguay and parts of Brazil) is in turmoil for several reasons: the collapse of Spanish empire and the political machinations of parties in new states, wars of genocide against native peoples, the massive influx of Africans living in bondage, and the meddling of naval and commercial powers (France and especially England) bent on establishing global trade networks. Civil wars and revolutions follow independence from Spain for roughly a decade. Uruguay, after war with Brazil, has obtained a fragile independence only three years old as the *Beagle* drops anchor.

Unexpectedly, the British frigate HMS *Druid*, at station in Montevideo, calls upon the *Beagle* to "clear for action" and "prepare to cover our boats." A military coup is brewing onshore, and four hundred horses owned by a British subject have been seized. The *Druid* readies six small craft of British marines and sailors armed to counter the insurgents. This show of strength prevents battle. Darwin affects an attitude of bemusement.[28]

A quest for Spanish records that might aid in navigating the Patagonian coast sends the *Beagle* scurrying days later across the Rio de la Plata to Buenos Aires. Greeting the arrivals as if they were buccaneers carrying cholera, a Buenos Aires guard ship opens fire on the *Beagle*. Her crew hears the "whistling of shot" over the rigging." Darwin comments, "We certainly are a most unquiet ship; peace flies before our steps." Preparing to return to Montevideo, Captain FitzRoy orders the canons loaded. "If she dared to fire a shot we would send our whole broadside into her rotten hulk."[29]

Fortunately, the *Beagle* never has to fire its canon; the jolt would no doubt have disturbed the timing of the ship's twenty-two chronometers. No other equipment is more essential to FitzRoy's navigational mission.[30]

Back in Montevideo, August 5, 1832, proves "an eventful day in the history of the *Beagle*. . . . [T]he Minister for the present military government came on board & begged for assistance against a serious insurrection of some black troops."[31] The chief of police informs FitzRoy that the mutinying soldiers have taken the citadel (which held the city's ammunition), broken open the jail, and armed the prisoners. Now the insurgents are threatening to ransack Montevideo, thus endangering the lives and property of Britons living there.

Americans who are in port occupy the Customs House. FitzRoy deploys fifty-two men, "armed to the teeth" with muskets, cutlasses, and pistols, Darwin among them. Heavily armed, they hoist out and man the *Beagle's* yawl, cutter, whaleboat, and gig. They garrison the principal fort, thus holding the mutineers in check, and wait for the cavalry.

> All through the evening they kept thundering away. Ball after ball flew over or fell short or kicked up the sand in the enclosure, but they had to fire so high that the shot fell dead and buried itself in the soft sand. We had no ricochet to fear, and though one popped in through the roof of the log-house and out again through the floor, we soon got used to that sort of horse-play and minded it no more than cricket.[32]

So speaks Jim Hawkins in *Treasure Island*. Echoing young Jim, Darwin, now the warrior scientist, two pistols at his side, rifle in hand, a magnifying glass in his pocket, defending the glory of England, would later reflect, "There certainly is a great deal of pleasure in the excitement of this sort of work.—quite sufficient to explain the reckless gayety with which sailors undertake even the most hazardous attacks."[33]

Probably to his disappointment, there is no exchange of gunfire; rebellion resembles cricket a bit too closely—more competition than bloodshed. The mustered company cooks beefsteak in the fort's courtyard while various parties attempt negotiations. Reinforcements surround the insurgents and force them to surrender. FitzRoy withdraws his troops, assured by government officials that although they have not been called into action, "the presence of these seamen certainly prevented bloodshed."[34]

Skirmishes nevertheless persist. At issue is a struggle between the military government of the Blancos (led by Manuel Oribe, a military hero of Uruguay's war for independence) and the elected government (Uruguay's first) of president Fructuoso Rivera, backed by the Colorados. At last a contingent of "1800 wild Gaucho cavalry," many of them Indians, enters the city—"a magnificent spectacle; the beauty of the horses & the wildness of their dresses & arms were very curious."[35] The constitutional government replaces the military governor, who flees. Darwin labels these comings and goings "imbecile changes."[36]

This strife in Montevideo is ongoing; lawlessness in Buenos Aires proves equally dangerous. The British play the role of international police, attempting to keep commerce thriving and shipping open around South America.

Upon a second crossing of the Rio de La Plata to Buenos Aires, the guard ship at the harbor entrance this time treats the *Beagle* with respect. Darwin disembarks for some shopping, noting the graceful walk and charming backs of the Spanish ladies, silk shawls folded around their shoulders, hair beautifully arranged and held high by an enormous comb.[37] Captain FitzRoy and Darwin call upon Donna Clara, also known as Mrs. Clarke, "an old decrepid woman: with a masculine face, & evidently even yet a most ferocious mind." Convicted of "some atrocious crime" and jailed at sea for transport, Donna Clara led her fellow convicts to murder the captain and most of the crew, pirate the ship, and sail to Buenos Aires. There she married a man of property, inherited his wealth, and earned affection for nursing wounded soldiers. Like Long John Silver, the outlaw Clara strikes an ambiguous posture: rapacious pirate and compassionate nurse. As to her perceived enemies, Donna Clara would "hang them all" or merely "cut their fingers off" for minor offenses. "The worthy old lady looks as if she would rather do it, than say so," remarks Darwin.[38] Her dreadful stories would indeed shock the plain folks of England.

On November 14, 1832, the *Beagle* anchors at Montevideo, where the officers attend a "Jubilee" grand ball honoring President Fructuoso Rivera, the same Colorado partisan restored to office with the assistance of the mustered *Beagle* crew.

Dousterswivel among the Gauchos

Following the exciting events in Montevideo and Buenos Aires, the *Beagle* sails south, surveying the desolate coast of eastern Patagonia. Two schooners help with this effort, one owned by an English trader, James Harris, who for a time serves as a pilot for the *Beagle*. On September 6, 1832, the *Beagle* and its crew arrive at Bahía Blanca (now a naval station), where they encounter some very rough-looking Spanish settlers accompanied by Indian prisoners impressed into serving the Spaniards.[39] As to the gauchos, Darwin records his first impressions:

> They formed by far the most savage picturesque group I ever beheld.— I should have fancied myself in the middle of Turkey by their dresses.— Round their waists they had bright coloured shawls forming a petticoat, beneath which were fringed drawers. Their boots were very singular, they are made from the hide of the hock joint of horses hind legs, so that it is a tube with a bend in it; this they put on fresh & thus drying on their legs is never removed.—The spurs are enormous, the rowels being from one to two inches long. . . . No painter ever imagined so wild a set of expressions.[40]

Led by a pair of these gauchos—land pirates incarnate—Darwin rides inland with Harris, who introduces him at the settlement as "un naturalista," a suspicious occupation loosely translated as "someone who knows everything."[41] As a consequence of ongoing war, the settlement has been frequently attacked. The locals suspect the *Beagle* party of being spies.

Darwin learns of the region's "barbarous and cruel warfare. . . . The Indians torture all their prisoners & the Spaniards shoot theirs. . . . [There is] a refinement in cruelty I never heard of": native children killing prisoners with nails and small knives.[42]

Soon Darwin reunites with the *Beagle* for its season of surveying Tierra del Fuego. Months later his encounters with gauchos resume north of the Rio de la Plata.

Thunderstorms and torrents of rain greet Darwin's arrival in Maldonado. Reports of deaths and extensive flooding are many. Guided by two well-armed men, he sets out on a geologizing expedition in search of marble and granite. They pass Laguna del Potrero where just the day before a man lay murdered, his throat cut. When gauchos meet the party at a small *pulpería* (a "drinking shop"), Darwin's compass proves mesmerizing. To his surprise, local people ask him "whether the earth or sun [moves]; whether it [is] hotter or colder to the north," as well as the location of Spain.[43]

Having impressed his audience with his mastery of geography, Darwin delights them by taking "Promethians" from his pocket and biting them between his teeth. These are matchsticks—explosives in miniature—in which combustible materials composed primarily of chlorate of potash (potassium chlorate), sugar, and camphor surround a tiny glass globule of sulfuric acid. The mixture is enclosed in paper; breaking the glass globule releases the acid to ignite the combustible mixture. The paper burns especially well if previously saturated in a solution of gunpowder.[44]

 Darwin's knowledge of venomous snakes, coupled with his obvious mastery of incendiary devices and geography, makes the people see in him "a good deal of the Dousterswivel"—a character from Sir Walter Scott's *Antiquary* described as "a tall, beetle-browed, awkward-built man, who entered upon scientific subjects . . . with more assurance than knowledge."[45]

Larger than matchsticks are "lightning sticks"—fulgurites to Darwin the geologist. He finds a host of these fascinating tubes of fused sand, forged by strikes of lightning, near Laguna del Potrero. Alas, he has found these novelties too late to woo Fanny Owen. Had Darwin only known, he might have shipped a few fulgurites back to England and impressed Fanny into calling off her engagement.

Moving on to Las Minas, Dousterswivel Darwin meets with "a great number of young Gauchos" who have come to the *pulpería* to drink and smoke cigars easily lit with Promethians:

> They are a singularly striking looking set of men.—generally tall, very handsome, but with a most proud, dissolute expression.—They wear their moustachios & long black hair curling down their necks.—With their bright coloured robes; great spurs clanking on their heels & a knife, stuck (& often used) as a dagger at their waists, they look a very different race of men from our working countrymen.—Their politeness is excessive, they never drink their spirits, without expecting you to taste it; but as they make their exceedingly good bow, they seem quite ready, if occasion offered, to cut your throat at the same time.[46]

Beneath a sky "brilliant with lightning" the *Beagle* gets under way to Rio Negro, over five hundred miles to the south along the coast of Patagonia. Darwin's "heart exults" at the thought "of all the glorious prospects of the future," especially doubling round Cape Horn once again in the coming summer.[47]

Having sailed far south, Captain FitzRoy deposits Darwin near the town of Carmen de Patagones. While the *Beagle* crew survey the coast northwards to Buenos Aires, Darwin journeys inland and north by horseback for three months with five gauchos, the Englishman Harris unable to join them. The ride starts at the foot of the ancient Walleechu tree, a sacred site to native people perched on a small promontory overlooking the pampas.

Decades of conflict with the indigenous Mapuche (also known as the Araucanians, or Araucanos by the Spaniards, a name that is now considered a derogatory term) have brought slaughter to the colonial inhabitants at the edge of the pampas and retaliatory genocidal attacks by settlers and soldiers. The region is one of the most hostile to Europeans in the Americas.

Shortly before Darwin's visit, General Juan Miguel de Rosas set out on la Campaña del Desierto (the Desert Campaign), leaving soldiers at fortified *postas* between the encampment of Carmen de Patagones and Buenos Aires. Rosas paid one tribe, the Tehuelches, to kill all other natives who crossed to the south of the Rio Negro or suffer annihilation themselves. According to Darwin, these fortifications have made travel "tolerably safe," and tolerably safe is what he needs to carry out duties as a *naturalista.*

Tolerably safe in the nineteenth-century pampas, nonetheless, does not exclude the possibility of being devoured by jaguars, murdered by gauchos, or speared with rhea-feathered bamboo lances, called *chuzos*, by the Mapuche who have escaped Rosas's Desert Campaign. Each evening his companions share stories of brushes with death. One recounts how, two months before Darwin's arrival, a group of warriors on horseback speared his two companions, then chased him down and entangled his horse's legs with a leather thong, forcing him to jump to the ground and cut the animal free. He sprang back into the saddle, dodging *chuzos* as he raced to the safety of the nearest *posta.*[48]

On August 23, 1833, news reaches Darwin that the "Indians had murdered every soul in one of the Postas. . . . [In response] the whole tribe was massacred."[49] Such barbarity prompts Darwin to reflect two weeks later, "Who would believe in this age in a Christian, civilized country that such atrocities were committed?" He finds the gauchos inferior "in every moral virtue" to the native peoples.[50]

As Darwin rides across the pampas with his gaucho guides, he revels in their wild lifestyle and tastes their culinary delicacies—steak from wild cattle, "ostrich" (actually rhea) eggs, puma tongue, and roast of armadillo. Regarding his long overland excursion, Darwin records his memory of the first night:

> There is high enjoyment in the independence of the Gaucho life—to be able at any moment to pull up your horse, and say, "Here we will pass the night." The death-like stillness of the plain, the dogs keeping watch, the gipsy-group of Gauchos making their beds round the fire, have left in my mind a strongly-marked picture of this first night, which will never be forgotten.[51]

In early September Darwin enters in his journal:

> The Indians are now so terrified that they offer no resistance. . . . [E]ach escapes as well as he can, neglecting even his wife & children.—The soldiers pursue & sabre every man.—Like wild animals however they fight to the last instant.—One Indian nearly cut off with his teeth the thumb of a soldier, allowing his own eye to be nearly pushed out of the socket.[52]

He comments on the ultimate prospects of General Rosas's campaign, "This war of extermination, although carried on with the most shocking barbarity, will certainly produce great benefits; it will at once throw open four or 500 miles in length of fine country for the produce of cattle."[53]

Shortly thereafter Darwin rides to the Sierra de la Ventana to ascend the Cerro Tres Picos. The climb is arduous and bitterly cold. His legs cramp; ice coats much of the jagged gray quartzite rock. The landscape is barren, the view across the pampas toward Bahía Blanca spectacular. Three years previously there was heavy fighting in these quartzite mountains.

A week later finds Darwin horrified to be eating a tasty, delicate dinner of *fetal* puma—a departure from armadillo steaks and "ostrich" eggs. He notes the prevalence of an invasive species, the common artichoke or thistle (cardoon). The dense, tall thistles provide an "excellent retreat and home for numerous robbers."[54] His discovery of several large arrowheads—presumed to be relics of pre-horseback hunting—convinces him that the horse was not native to South America.[55]

Upon his return to Buenos Aires, he stumbles upon another rebellion and finds the city blockaded by Rosas's men, who believe that "by stopping the supply of meat they will certainly be victorious."[56] His passport being from Rosas, Darwin gains entrance to the city, then leaves for collecting north to the Rio Paraná by horse to study the natural history of the region for the next two months.

After crossing to Montevideo on the north shore of the broad estuary of the Rio de la Plata, Darwin departs again, guided by a *vaqueano*, to travel the north shore of Rio de la Plata to Colonia del Sacramento. He reports being in constant danger of jaguars. During this venture "the dexterity with which some Peons crossed over the rivers" amuses Darwin:

> As soon as the horse is out of its depth, the man slips backwards & seizing the tail is towed across; on the other side, he pulls himself on again.—A naked man on a naked horse is a very fine spectacle; I had no idea how well the two animals suited each other: as the Peons were galloping about they reminded me of the Elgin marbles.[57]

On Darwin's return to Montevideo, a drunken man calls him a "Gallego," an expression synonymous with saying he is worth murdering. The modern meaning of the term is "your mother is a prostitute," the worst possible of insults requiring a knife fight, sometimes to the death. "The only manner of fighting amongst the Gauchos [is] stabbing each other," writes Darwin.[58] Darwin ignores the affront and avoids a nasty fight.

Immeasurable Treasure

The end of the year 1833 brings to a close Darwin's coming-of-age escapades riding in the company of gauchos, the pirates of the pampas. In many respects, like young Jim Hawkins, Darwin no longer wants anything to do with such wicked and cruel men and their "dreadful stories." He has matured and, with maturity, come to appreciate nature's bounty at least as much as the romance of adventure. He and the *Beagle's* crew must turn their attention to the southernmost extremities of South America. In 1834 they will pass through the Strait of Magellan to begin charting Chilean waters.

Incidental to the *Beagle*'s nautical mapping tasks, to him his collections now matter most. South of Patagonia in Tierra del Fuego, north along the Chilean coast, landward through Andean passes, and island-hopping westward across the Pacific lie more secrets to unbury. Geologic discoveries will join the horde of specimens and fossil treasures from the Rio de la Plata and its tributaries, Bahía Blanca, Maldonado, and Brazil—the menagerie that Philos has stowed away for English naturalists to examine later: a pirate-naturalist's treasure trove.

Gauchos, vaqueros, pumas, jaguars, storms, pampas, deserts, and mountains have tested young Charles Darwin. Malaria and a hunting accident have cost the lives of several compatriots; an infected wound and stomach illness, not to mention intolerable seasickness, have challenged his health. Darwin has witnessed barbarism and cruelty, slavery and the repatriation of native hostages, urban insurrection and frontier warfare. Yet despite the tribulations, he remains buoyant and enthusiastic. His *Beagle* voyage will bequeath to humanity the origin of origins stories, an immeasurable treasure, no matter how shocking to "plain country people" his ideas might become.

Following a year of mapping the channels of the Chonos Archipelago and the harbors and headlands of Chile's Pacific coast, the *Beagle* sets sail on September 7, 1835, for the Galápagos Islands and legend.

3

FOSSILS, FOLLY & FAULTS

Rounding South America

Nothing can be more improving to a young naturalist than a journey in distant countries. . . . I have too deeply enjoyed the voyage not to recommend to any naturalist to take all chances, and to start on travels by land if possible, if otherwise on a long voyage. He may feel assured he will meet with no difficulties or dangers (excepting in rare cases) nearly so bad as he beforehand imagined.

Charles Darwin, Beagle *Diary*

Dangers "so bad as beforehand he imagined" may fail to greet Darwin, but catastrophes are no stranger to his journey as he sails the southern oceans and treks the Andes Mountains. He endures the unrelenting fierceness of Cape Horn storms, the devastation of earthquake and tsunami, and the debilitating effects of disease. Nevertheless, upon returning home, he recommends to any naturalist "to take all chances" and seek improvement by journeying to distant countries.

Against a background of strife and danger, he seeks creatures new to him—and often to science—whether the inhabitants of the land or the buried denizens of the past, the victims of extinction. He certainly follows his own advice in pursuing his primary avocation as a "naturalista." Whether riding with General Rosas's gauchos or charting his own course, Darwin spends long stretches of time inland. Meanwhile, the *Beagle* conducts its mapping duties or lies refitting in port. On each of his excursions the peripatetic Philos revels in a cornucopia of fossil and faunal discoveries.

Bahía Blanca and Santa Fe

Having been put ashore on the Atlantic coast south of Buenos Aires, Darwin finds himself digging his way around Bahía Blanca's Punta Alta and the nearby Monte Hermoso. Here lies fossil nirvana: bleached bones protrude from clayey cliffs. On two occasions Darwin's visits to this site reveal the remains of enormous creatures, some previously undiscovered, and some containing organic matter. The specimens are truly extraordinary. He digs out a skull of *Scelidotherium* (a genus of giant ground sloth) but thinks at first it is that of a rhino. He finds a *Megatherium* (another giant ground sloth) tooth within a jawbone. Ossified polygonal plates, later identified as coming from a carapace of the extinct glyptodont *Hoplophorus* (a giant armadillo), intrigue him.[1]

Remarkably, a third genus of ground sloth emerges from these cliffs: *Mylodon*, whose fearsome descendant the mapinguari is rumored to haunt the deepest recesses of Amazonia even today.[2] Remains of the monstrous sloth no longer seem quite so novel as they did when he first began collecting.

Darwin finds fragments of at least six distinct beasts, including one he presumes to be, as would be known to the Dread Pirate Roberts in *The Princess Bride*, a rodent of unusual size (ROUS).[3] It has the curved teeth that appear to mark it as a gnawing animal. He surmises that the fossil ROUS could be the progenitor of the living capybara, or "river hog," a wild giant cousin of the guinea pig. Darwin's rhino-sized rodent fossil, however, turns out to be neither rhino nor rodent. The enormous "gnawer" belongs to an extinct creature from an extinct order of South American mammals, the notoungulates. The mysterious beast eventually receives the appellation *Toxodon*.

He stocks his collections for shipment to England with skins and skeletons from living creatures as well as fossils of extinct megabeasts. His treasured bones include those from ancient rodent relatives of capybaras, agoutis, and tuco-tucos. Darwin counts among his onboard menagerie a number of lizards and frogs, at least a half-dozen species of snakes, and a great many species of birds, among them oven bird, woodcock, shrike, black-hooded oriole, kingfisher, finch, godwit, warbler, parrot, thrush, stilt, hornero, partridge, woodcreeper, flycatcher, bunting, rail, sandpiper, woodpecker, pipit, tit, plover, skimmer, owl, vulture, dove, tyrant flycatcher, water hen, oystercatcher, and white tern.[4] There are dung beetles and water fleas (*Daphnia*), flatworms and fish, and couple of armadillos, all destined for service to science.

Guinea pigs, modern cavimorph rodents farmed as pets and food, are abundant on the pampas; so too is the nasal-signaling rodent that chirps its name, the tuco-tuco. The tuco-tuco lives in burrows marked by small mounds. Sporting bright yellow teeth, it churns the ground so thoroughly that horses' hooves sink in. Its name derives from its repetition of a nasal noise four times in succession: "tu-co-tu-co." Tucotucos "are stupid in making attempts to escape"; many are blind.[5]

Darwin is neither the first nor the last to find fossilized footprints around Bahía Blanca. Recent trackway discoveries at Monte Hermoso feature footprints made by human children—probably foraging for flamingo eggs—about seven thousand years ago.[6]

The terraces of Bahía Blanca display even older tracks of three-toed *Rhea americana* (the ostrich-like bird of South America), *Macrauchenia* (the snouted,

Figure 3.1 The tuco-tuco (*Ctenodactylus*), a burrowing rodent that chirps its own name. Illustration by Jan Glenn.

camel-like three-toed ungulate from an extinct group restricted to South America: the Litopterna), and *Palaeolama*, a two-toed ancestor of the guanaco, or, as referred to by Darwin's term, the American camel. Radiocarbon dating of material from nearby Pehuen-Có dates these fossils from twelve thousand to sixteen thousand years ago.[7]

Guided by gauchos, and having returned to Buenos Aires, Darwin roams north to Santa Fe. He pauses to excavate a *Toxodon* skeleton, the rhino-sized creature he persistently takes for a rodent. He observes deer and "ostriches" (rheas) killed during a recent severe thunderstorm by hailstones the size of apples. In Santa Fe he comes across massive bones embedded in a cliff and mistakenly assumes they are from a mastodon. Because of their position within the cliff, the locals interpret these bones as the remains of a giant burrowing animal. Darwin's guess was closer: the defunct creature was indeed a proboscidean (elephants and related families) in the genus now known as *Stegodon*.

Hearing of some giant bones nearby, Darwin rides to another site and extracts a few, which "as usual" turn out to be those of the giant ground sloth *Megatherium*. En route back to Montevideo he purchases a "very perfect" part of a *Megatherium* skull "for a few shillings."[8] He totes his cargo of skulls, teeth, and femurs back to the *Beagle* and anticipates the next leg of his "journey of researches," the phrase that will introduce his first book.

Patagonian Coast and Tierra del Fuego

As the *Beagle* nears Tierra del Fuego in late November 1832, the repatriation of FitzRoy's hostages moves ahead according to plan. Three have survived; one—Boat Memory—died of smallpox in England. The ship makes rendezvous at the bay of San Blas with two schooners that will assist in surveying the coast. Off the bay, butterflies flitter as far as the telescope can reach. The crew shouts, "It is snowing butterflies!"[9]

Approaching the southern tip of South America, FitzRoy's *Beagle* zigs and zags its way through the same ends-of-the-earth waters Magellan so expertly navigated three centuries earlier. Place-names given by the Portuguese explorer echo from the shores: Patagonia and Tierra del Fuego. As Magellan navigated the straits through this archipelago, the indigenous peoples lit fires—hence "Land of Fires," or "Tierra del Fuego." Supposedly "patagon" translates as "bigfoot" and refers not to human feet exactly but to the fur boots worn by the people Magellan met at Port St. Julian—including those he took hostage as curios.[10]

Smoke signals the *Beagle*'s arrival to the south of Cape San Sebastián, just days after the "heaviest squall" Darwin has ever known. It is a harbinger of the crew's near future. They glimpse men of the Eastern Ona tribe, feared by FitzRoy's captive Fuegians, whose native Yaghan people dwell in a nearby region. On December 17 the *Beagle* passes through the Le Maire Strait and pot-boiling seas to anchor in Good Success Bay. "Perched on a wild peak overhanging the sea [Fuegians spring up] & waving their cloaks of skins [send] forth a loud sonorous shout."[11] Two days later Darwin "attempts to penetrate some distance into the country" and finds "hills . . . so thickly clothed with wood as to be quite impassible." He manages to ascend through a forest of evergreen beech festooned with moss and lichen. The ground is immersed in dank decaying matter, making "death instead of life the predominant spirit."[12] The *Beagle* sits below at anchor; two guanaco stand close by, alpine in habit and nimble on the rocks.

December 22, the first day of summer, 1832, finds the *Beagle* traversing stormy southern latitudes. That night, Darwin records, as the ship rounds the tip of Deceit Island, Cape Horn sends an ominous gale "right in our teeth." As the ship sails close-reefed the next day, its westward progress slows. On December 24 the crew can still view Cape Horn "veiled in . . . mist . . . its dim outline surrounded by a storm of wind & water: Great black clouds . . . [roll] across the sky & squalls of rain & hail [sweep] by us with very great violence." The ship runs for safe harbor at Wigwam Cove and finds smooth water for anchorage, as the men welcome a day of layover for Christmas at the antipode of hearth and home. Here the sea submerges the southern terminus of the Andes to create a rugged chain of islands, their ragged peaks towering above the surf. Every several minutes strong cold breezes—"Whyllywaws"—descend from the steep slopes to remind the crew of the fierce gales beyond their cove.[13] Winter weather delayed departure from England a year earlier, and violent flogging followed the *Beagle* crew's bacchanalian Christmas behavior in 1831. This is summer in the Southern Hemisphere, and the weather is even stormier. Discipline has been tightened for sailing the treacherous waters of Cape Horn. Surveying the channels of Tierra del Fuego and the coast of Patagonia must be synchronized with summer in the southern ocean. Even so, gales whip up the most frenzied seas.

For several days the "weather is so bleak & raw as to render boating rather disagreeable," notes Darwin, and although it is summer and the latitude is a mirror of Edinburgh's, "the climate is singularly uncongenial": "Much rain & violent squalls." "Tremendous gale . . . continued rain & wind." "The sun's rays seldom have much power." Summiting at 1,700 feet on a Christmas Day hike, he observes the "nakedness of the land," interrupted cove by cove with conical four-foot-high wigwams made of hay, "formed of a few branches & imperfectly thatched with grass, rushes &c." The wigwams of the peoples living at the southern extremity of South America are mimics in miniature of the drowned Andean peaks. Loggerheaded steamer ducks swim among the coves, using their wings as paddles. Penguins abound.[14]

Neither a temperate Scottish summer nor a cheery English Christmas, the antipodean holiday season promises nothing but the harshest of sailing conditions. No one could imagine how harsh.

An Encounter with a Graybeard

The Drake Passage lies south from Cape Horn, stretching to the South Shetland Islands of Antarctica. Legendary waves traverse these waters. They are known to mariners as "graybeards." Sailors claim that from crest to crest a graybeard may span a mile; from trough to crest a graybeard may rise two hundred feet, though in reality a rise of one hundred feet is more accurate.[15] The mariners' stories, however, are not entirely apocryphal.

Between the latitudes of 60°S and 65°S a low-pressure trough rings Antarctica, its intense drop in pressure propelling storm clouds upwards and drawing in westerly winds over vast fetches of the ocean. These Westerlies blow unimpeded over frigid waters as storms form, dissipate, and re-form in all seasons of the year. In the summer, as heat moves poleward, the winds at 55°S—Cape Horn lies just one degree farther to the south—become intense. No place at sea inspires deeper dread, and nowhere else are gales so frequent.[16] Gusts of wind attain hurricane force, driving immense seas before them:

> I was mesmerized by the seas—I'd never seen anything like them before. On most waves, the boat would ride up and the swell would pass under the boat and break later. Every now and then a larger one would come, maybe 40 to 50 feet, with the classic "graybeard appearance." The top 3 to 6 feet would be foamy streaming grayish water, churned and blown off by the wind. Below this was the dangerous stuff, a wall of water that could break

> the boat or crush you. As I watched, I could see the surface layer of this wall of water sort of break loose and slide down the face with a glistening, white rippling effect, resembling water sliding down a water slide in an amusement park ride. Finally one of these waves broke on the boat and we began the down-wave slide again.[17]

The *Beagle* puts to sea on New Year's Eve day and sails day and night against the Westerlies. Currents work against the ship's progress. Beating against strong gales, ocean current, constant rain, and heavy seas, it traverses barely a mile and a half in four days. After almost ten days the ship doubles round Cape Horn again, having made virtually no progress toward its Christmas Sound destination in the west-central portion of the archipelago and the proposed mission site for the Fuegians. "Scarcely for an hour" has Darwin "been quite free from sea-sickness." Immersed in miseries and discomfort caused by the ship's severe pitching, he feels he cannot "hold out much longer."[18]

After some days at last of fair winds, the *Beagle* finds itself within easy sighting distance of Christmas Sound and the prominent peak, York Minster, the inspiration for the name given to one of FitzRoy's captives captured nearby. Alas, a violent gale forces the *Beagle* to "shorten sail & stand out to sea." A graybeard breaks, spewing spray "over a precipice" at least two hundred feet high.[19] The weather worsens and the seas gain in treachery.

Storms test the crew's endurance the day after and the next. The lookout man strains to watch the horizon, for the ship's officers are no longer able to fix its position. Incessantly, the winds drive the ocean spray, obscuring vision. The "ominous" sea has churned up "so much foam that it [resembles] a dreary plain covered by patches of drifted snow."[20] The sea is a forest of graybeards. As the crew labors, Darwin observes an albatross glide effortlessly upwind, as if to mock the ship's struggle.

Sunday, January 13, 1833. Noon. The *Beagle* flirts with doom, almost drowning any chance of Darwin's journals, notes, and voyaging leading to his theory of origins. A series of three massive waves approaches. FitzRoy's seamanship masters the first. The *Beagle* climbs upwards and crests its summit, but loses its momentum through the water and thus the ability to maneuver. "The second wave [deadens] her way completely."[21] The third, a "great sea," washes overboard, breaking loose a whaleboat from its tackle. Water reaches the poop and forecastle cabins and does damage to Darwin's collections.

The crew cuts away the last lines holding the thrashing whaleboat and it slips into the sea. This moment has its irony: theft of a whaleboat drove FitzRoy to take hostages to trade for its return during his first *Beagle* voyage. Now, on its second voyage, a whaleboat abandons ship.

Figure 3.2 Storm-tossed and soaked collections. Illustration by Jan Glenn.

Washing over the *Beagle*, the wave fills the "decks so deep, that if another had followed it is not difficult to guess the result," records Darwin. Decks awash, water pressing on its bulwarks, the ship flounders on her lee side, nearly capsizing in the roiling sea. "At last the ports were knocked open & she again rose buoyant to the sea."[22] FitzRoy later writes: "Had another sea then struck her, the little ship might have been numbered among the many of her class which have disappeared. . . . The roller which hove us almost on our beam ends, was the highest and most hollow that I have seen, excepting one in the Bay of Biscay, and one in the Southern Atlantic."[23]

The next day Darwin laments:

> We stood to the North to find an harbor; but after a wearying search in a large bay did not succeed. I find I have suffered an irreparable loss from yesterdays disaster, in my drying paper & Plants being wetted with salt-water.—Nothing resists the force of an heavy sea; it forces open doors & sky light, & spreads universal damage.—None but those who have tried it, now the miseries of a really heavy gale of wind.—May Providence keep the *Beagle* out of them.[24]

For a French whaler moored at the nearby East Falkland Island's Port Louis, the storm proves more disastrous. The hurricane-force winds of the mid-January storm has ripped the ship from anchor and driven it into the shore, crushing its hull. FitzRoy purchases many of its stores and supplies and promises its marooned crew of twenty-two sailors passage back to Maldonado in Uruguay.

The *Beagle* has narrowly escaped the jaws of catastrophe. The sea fails to swallow Darwin, his journals, and collections. Given barely twelve miles of progress in three weeks, Captain FitzRoy decides to settle Jemmy Button, Fuegia Basket, and York Minster, and the Reverend Richard Matthews, age twenty-two and "of an eccentric character," at Woollya, Jemmy's home cove in the eastern straits of the Tierra del Fuegan archipelago.[25]

A Plundered Mission

Whalers carry sensible stores; missionaries often do not. In addition to the Reverend Matthews's eccentric character, the sponsors of the mission had prepared for it a "choice of articles [that] showed the most culpable folly & negligence," writes Darwin. "Wine glasses, butter-bolts, tea-trays, soup turins, mahogany dressing case, fine white linen, beavor hats & an endless variety of similar things shows how little was thought about the country where they were going to. The means absolutely wasted on such things could have purchased an immense stock of really useful articles."[26]

FitzRoy misunderstands the name of the local people to be the "Yapoo Tekeenica." Thomas Bridges, a later missionary and translator of Fuegian dialects, corrects this misunderstanding. "Yapoo" means "otter"; "Tekeenica" means "I do not understand you" to the Yahgashagalumoala tribe (literally, "People from Mountain Valley Channel," aka, the Murray Narrows).[27] Convention shortens this name to "Yaghan," Jemmy Button's people. Contracting the tribal name proves prescient; by the mid-twentieth century they number only dozens.

"Cleared and rich" ground at Woollya promises to grow a nourishing "kitchen garden" of potatoes, carrots, turnips, beans, peas, lettuce, onions, leeks, and cabbages—providing fodder for the soup tureens.[28] Perhaps Captain FitzRoy has assumed that Fuegians lack nothing so much as healthy European vegetables to augment their shellfish diet.

Jemmy's extended family, painted in white, red, and black, appear by canoe together with many others expecting gifts. York Minster takes Fuegia Basket as his wife. Within a few days the contingent from the *Beagle* judges the settlement well established and departs. The yawl and one whaleboat return to the ship; Darwin and his captain go exploring in the other two whaleboats.

Whales accompany them as they row through the deep but narrow Beagle Channel—christened during the ship's earlier voyage—ringed by drowned Andean peaks draped in glacial ice. A change in weather leaves Darwin and his compatriots sunburned. At one point Fuegians arrive to threaten the evening's camp. The band moves to another site, realizing that the courage of the Fuegians "is like that of a wild beast" and that each one "would endeavor to bash your brains out with a stone, as a tiger would be certain under similar circumstances to tear you." The "similar circumstances" being, of course, territorial trespass. "The occasional distant bark of a dog reminds one that the Fuegians may be prowling, close to the tents, ready for a fatal rush," Darwin enters in his journal.[29]

Glaciers extend from the mountains to the water. Darwin can imagine nothing "more beautiful than the beryl blue of these glaciers"[30] and enjoys evening after evening of boat camping, snugly bedded in his sea cloak that by each dawn is frozen solid.[31]

With boats beached one afternoon to take lunch, the picnicking party admires "the beautiful colour of [the] vertical and overhanging face" of a nearby glacier. The beryl blue belies an unpredictable hazard suggested by the icebergs drifting menacingly through the Beagle Channel. Without warning the frozen face gives thunderous way to calve a massive iceberg, the sound echoing throughout the channel. More dangerously, "a great wave [rushes] onwards" as the sea "[heaves] up in a great heap of foam" that tosses the boats "along the beach like empty calabashes." Darwin and the other seamen act quickly to secure these "calabashes," so utterly essential to survival. He comments, "If they had been washed away; how dangerous would our lot have been, surrounded on all sides by hostile Savages & deprived of all provisions."[32]

The following day, Postillion-Philos, the hero naturalista, one of many dread English pirates to the Fuegians, and a Dousterswivel to the gauchos, earns lasting geographical recognition: FitzRoy names a sound and overlooking mountain after his gentleman companion. Mount Darwin peaks at well over a mile above Darwin Sound.

While Darwin and his mates are busy securing calabashes and replacing Yahgashagalumoalan names for local landmarks with proper English ones, native people steal Reverend Matthews's soup tureens, beaver hats, wineglasses, and fine linens, the articles of "folly and negligence" graciously provided by the Church Missionary Society and accepted by Captain FitzRoy. No china remains for serving soup according to proper British etiquette. More than three hundred Fuegians come and go to watch the curious happenings at the mission and English vegetable garden and, from the Englishmen's perspective, to help themselves to whatever bounty they can. Perhaps they feel entitled to compensation for land seized and occupied without permission.

On February 5, the weather properly restored to the status of "miserable," the exploring party returns to Woollya. En route they encounter signs of trouble: a native woman wearing one of Fuegia Basket's fine English garments. She is not Fuegia Basket. Upon reaching the beach at Woollya, they encounter a fleet of canoes and several dozen Fuegians "much painted, and ornamented with rags of English clothing."[33] The Reverend Matthews, feeling murder imminent, accepts inevitable retreat, abandons the mission, and rejoins the *Beagle*. A few years later he will resume missionary work in New Zealand.

Jemmy Button, York Minster, and Fuegia Basket remain at Woollya. Following a week of surveying, FitzRoy returns to check on his charges, finding tranquility and a few remaining vegetables. Darwin considers the prospect of leaving the Fuegians "amongst their barbarous countrymen" to be "quite melancholy" and expects them to return shortly to their "uncivilized habits."[34] The savage habits that next capture Darwin's attention are those of Europeans contesting for control of the Falkland Islands.

Guerras de Las Malvinas

In the Falklands, Darwin delights in the waddling penguins and ducks that paddle with their wings and croak like bullfrogs (the loggerheaded or steamer duck). He finds their heads to be quite strong and almost unbreakable when struck by his geological hammer. Their beaks are equally tough. From steamer duck dung, Darwin realizes the usefulness of these hardheaded traits: the ducks feed on shellfish.[35] They probably would have preferred that Darwin stick to hammering on rocks.

In 1833 the Falklands were ideally positioned to service Southern Ocean whaling fleets—the likely reason why the islands were contested. Upon the ship's reaching the harbor of Port Louis, a French settlement on East Falkland Island, news "that England had taken possession of the Falkland Islands" stuns Darwin. He is aware of Buenos Aires' claims to the islands (Las Malvinas) and of British interests from decades earlier based on the establishment of a colony at Port Egmont. The two powers nearly come to war while Darwin goes about clobbering steamer ducks. They finally do so in 1982.[36]

Darwin comments: "The present inhabitants consist of one Englishman . . . 20 Spaniards & three women, two of whom are negresses —[T]here are about 5000 wild oxen, many horses, & pigs."[37] The lone Briton, a Mr. Dixon, is a virtual emperor of the oxen.

While the *Beagle* is anchored in the Falklands, a storm swamps the ship's yawl. FitzRoy purchases an American sealing vessel from a Captain Lowe, whom Darwin characterizes in an understatement as "a notorious & singular man who has frequented these seas for many years & been the terror to all small vessels." The description translates as "pirate." Captain Lowe is a person of ambiguous morals: he has onboard American sealers rescued from a storm-wrecked ship. He also has onboard a piratical mate, a slaver who saw action against the HMS *Black Joke*. Later, Captain FitzRoy places the mate under arrest.[38]

Summer season's surveying has ended. Autumn—April—finds the ship passing the fossiliferous cliffs at the mouth of the Rio Negro, "El Dorado to a geologist," amidst "celestial" and quite "cloudless skies, light breeze & smooth water." Plans are set to make sail for Maldonado in Uruguay, where "things have been going on pretty quietly," remarks Darwin, "with the exception of a few revolutions."[39]

After refitting the *Beagle* in Montevideo (thus giving Philos the months needed to gather his inland menagerie), FitzRoy declares that the time for mapping the channels near Cape Horn has returned, for summer is about to begin. Conrad Martens replaces the ship's artist, Augustus Earle, who has fallen quite ill. The *Beagle* heads south on December 7, 1833.

Seventeen days later the crew makes anchor at Port Desire (Puerto Deseado), where Darwin notes a "great *level* plain" extending in every direction and "divided by vallies."[40] Oyster and mussels beds on this dry, barren plain prove to Darwin its uplifted past, for once these strata must have been below sea level. Barely fossilized and appearing virtually the same as living species (the mussel shells retain their blue color), they stand perched 247 feet above sea level. Such observations vindicate Charles Lyell's geology—extrapolating the geologic processes of the present to interpret the ancient earth. In the new year Darwin will experience directly the actual cause of uplift operating in the present.

Good hunting means fresh guanaco meat for Christmas dinner.

> After dining in the Gun-room, the officers & almost every man in the ship went on shore.—The Captain distributed prizes to the best runners, leapers, wrestlers.—These Olympic games were very amusing; it was quite delightful to see with what school-boy eagerness the seamen enjoyed them: old men with long beards & young men without any were playing like so many children.—certainly a much better way of passing Christmas day than the usual one, of every seaman getting as drunk as he possibly can.[41]

"Slinging the monkey" entertains the sailors at Port Desire, as captured in a Martens watercolor. A sailor hangs upside down, blindfolded, holding a stick, swinging in the sling. The others surrounding the tripod attempt to strike the poor victim, who returns the blows blindly in a vicious game of "tag, you're it." "It" means taking a turn as the monkey. "FitzRoy thought that this was quite exciting and motivating to divert the men to use a live person in slinging the monkey."[42]

The day after Christmas 1833, Darwin hikes five miles south to cliffs of the "same great oyster bed . . . being upheaved." He matter-of-factly comments on this outcrop, strong evidence favoring a revolution in geological thought, as "the usual geological story." Two days later, still on board the yawl, Darwin and a small party venture upstream to a plain of "sandy chalk, & gravel," where "all is stillness & desolation." On New Year's Day a walk up a distant hill reveals an Indian grave, the remains of campfires, and several horses' bones. Others join Darwin to "ransack the Indian grave in hopes of finding some antiquarian remains," but to no avail.[43]

For some time Darwin has heard of a small "ostrich" (actually a rhea) said to inhabit only the southern reaches of Patagonia. Eager to add this variety to his collections, Darwin settles down to enjoy a summer dinner of "ostrich" shot by the ship's artist, Conrad Martens. At first thinking the tender meat is that of a youngster, he soon realizes that in his hands is the elusive and diminutive species he seeks.[44] He scavenges the remaining victuals in order to return the samples to England, where they become the type specimen, named in his honor, *Rhea darwinii*. (*Rhea pennata* is now the accepted name, based on identification prior to Darwin's.)

While leaving Port Desire for Port St. Julian 110 miles to the south, the *Beagle* strikes a submerged rock but does not suffer serious damage. At Port St. Julian, Darwin geologizes upriver from the bay, noticing terraces and sediments shed by the Andes. Here he discovers a fossil *Macrauchenia* (the three-toed, trunk-snouted, camel-like, horse-sized herbivore), though he misinterprets this find as a mastodon skeleton. (The expert anatomist and Richard Owen—coiner of the term "dinosauria"—later corrects him.) Lacking knowledge of the immigration of early camel family creatures to South America during Pleistocene times, Darwin eventually and erroneously concludes that *Macrauchenia* is ancestral to the wild guanaco and domesticated llama. He surmises that the geographic affinity of fossil and modern creatures implies descent—a reasonable principle, but in this instance incorrect.[45]

Thirst overcomes FitzRoy and several companions during a search for "pozos de agua dulce" (freshwater ponds). Parched but still strong and able, "accustomed . . . to long excursions on shore," Darwin hikes on, only to find dried salt flats; others eventually rescue the party with water brought from the ship.[46]

A gravel plain studded with large oyster shells extends hundreds of miles along the southern coast of Patagonia, reaching Tierra del Fuego, telling "a story of former times with almost a living tongue."[47] Terraced gravels, sorted and unsorted, beds of seafloor sediments, and shells suggest a land lifting and subsiding several hundreds of feet multiple times.

The scenery reminds Darwin of the treacherous attempts to round Cape Horn, when, thwarted by incessant storms, and despite his legendary seamanship, Captain FitzRoy chose to set the Fuegians and their missionary companion ashore at Woollya. Its protected bay lies well east of the islands Fuegia Basket and York Minster call home. Now Darwin and the *Beagle* return to Woollya for a final visit.

Fuegian Demise

On March 5, 1834, the *Beagle* once again anchors at Woollya in Ponsonby Sound, off the Beagle Channel. FitzRoy intends to leave Tierra del Fuego for good and continue his charting along the coast of Chile. The Woollya layover is to bid farewell to Jemmy Button.

They find him nearly unrecognizable. By this time the Ona tribe (the Sel'nam or Onawo people) as well as York Minster have stolen everything from Jemmy; York Minster and Fuegia Basket have fled west. No one knows what has become of the soup tureens.

When left earlier, Jemmy was "fat" and "particular about his clothes . . . afraid even of dirtying his shoes; scarcely ever without gloves & his hair neatly cut." Darwin "has never [seen] so complete and grievous a change." Now Jemmy Button is "thin pale & without a remnant of clothes, excepting a bit of blanket round his waist: his hair, hanging over his shoulders; & so ashamed of himself."[48]

Nothing fascinates Darwin more than the Fuegians migrating between their own culture and their English affectations. He notes their reversion to Fuegian norms of dress, gesticulation, and diet. He remarks on their exceptional talent for learning languages as well as their extraordinarily perceptive vision.

FitzRoy has wrongly presumed that their exposure to European custom, religion, and language will inevitably dispel superstition, usher in progress, and, not coincidentally, serve the interests of British commerce. He believes so strongly in this transformation that he perceives changes in the shapes of the Fuegians' foreheads due to England's "civilizing" influences.[49]

Darwin finds himself much less sanguine about the marks of Englishness sticking to Jemmy like a star on a Sneetch. Nor does he accept that face, forehead, and head bumps reveal secrets of character, personality, or racial disposition—the stock beliefs of phrenologists and physiognomists such as FitzRoy.[50] A change in attire and a new tongue to speak make no difference in cranial morphology. Darwin attributes Jemmy's nature not to an inherent racial disposition but to the overriding effects of immersion as a child in Fuegian culture and environment. Darwin glimpses in Jemmy the lifestyle of ancient Britons, untransformed by European civilization's morals. Darwin adheres to a belief in the common origin of humanity, whether Briton, Ona, Yaghan, or Alacaloof. For him, education accounts for and bridges the differences.

Darwin writes with evident affection for Jemmy and equal astonishment at his partial transformation. The days of farewell to the Fuegians lead him to enter in his journal on February 25, 1834, a telling description:

> Their country is a broken mass of wild rocks, lofty hills & useless forests, & these are viewed through mists & endless storms. In search of food they move from spot to spot, & so steep is the coast, this must be done in wretched canoes.—. . .Their skill, like the instinct of animals is not improved by experience; the canoe, their most ingenious work, poor as it may be, we know has remained the same for the last 300 years.
>
> Although essentially the same creature, how little must the mind of one of these beings resemble that of an educated man. What a scale of improvement

> is comprehended between the faculties of a Fuegian savage & a Sir Isaac Newton—Whence have these people come? . . . [W]e may therefore be sure that he enjoys a sufficient share of happiness (whatever its kind may be) to render life worth having. Nature, by making habit omnipotent, has fitted the Fuegian to the climate & productions of his country.[51]

In these words are the serious reflections of a young man in search of self, of standing as a naturalist, and of understanding of the human condition in the world. He attributes the state of the Fuegians not to descent into savagery caused by original sin but rather to the inhospitable lands and climate of Tierra del Fuego. Climate shapes land, plants, animals, and people. With time, by virtue of applying the "mental faculties" to the "social instincts"—each a product of natural selection—progress may ensue (in Darwin's mind leading, of course, to becoming more like Englishmen) and bridge the distance between Jemmy Button and Sir Isaac Newton.

Through the writing of *On the Origin of Species*, Darwin will remain silent on the role of natural selection in creating *Homo sapiens*. Publication of his ideas on human descent, no doubt shaped by his years circumnavigating the globe in his twenties, must wait until 1871.[52]

The laws of nature, as Darwin espouses, do not cause the misery of the poor, the squalid state of much of humanity, or the brutal hatred of one people for another. He sees in Jemmy nothing other than his own distant past.

Jemmy ultimately dies of measles in 1864. York Minster is slain in a violent act of revenge for his killing of an Alacaloof (Fuegia Basket's people), circa 1840. While in her fifties, Fuegia takes a new husband, eighteen years old. She dies in her sixties, most likely in the year 1883, shadowed by suspicion of working as a prostitute from time to time onboard English ships. Tierra del Fuego's original inhabitants are now "virtually extinct."[53]

Decades after Darwin's voyage, the Patagonian Missionary Society tries again to establish a mission in Tierra del Fuego; all but one member from their ship, the *Allen Gardiner*, perish in a massacre with Jemmy Button in a suspicious role. The year is 1859, the same year as the publication of *On the Origin of Species*. Within a generation, disease depopulates the area, and eventually the Argentine government turns the mission site at Ushuaia in the Beagle Channel into a prison.[54]In 1834, though, these dire endings still lie in the future. The *Beagle* departs eastward, the culture of Tierra del Fuego's canoe nomads poorly understood.

With no wish to return to England, Jemmy happily makes a gift of two otter skins. Darwin records in his journal:

> He seems to have taught all his friends some English. . . . Every soul on board was as sorry to shake hands with poor Jemmy for the last time, as we

> were glad to have seen him.—I hope & have little doubt he will be as happy as if he had never left his country. . . . He lighted a farewell signal fire as the ship stood out of Ponsonby Sound, on her course to East Falkland Island.[55]

Volcanoes and Earthquakes

Following a second visit to the Falkland Islands (once more in the midst of a revolt—this time subdued by the HMS *Challenger*), the *Beagle* spends April and May 1834 in southern Patagonia at the mouth of the Rio Santa Cruz. While the ship is dry-docked for repairs to its copper hull, Darwin and a party attempt an upriver excursion in order to reach the Andes. A series of terraces captures his attention as well as a deep basalt canyon. He observes twenty condors at once. Along the way he shoots a condor with an eight-and-one-half-foot wingspan. Darwin interprets the river valley as an uplifted seafloor, formerly a strait between oceans. As they near Lake Argentino, a decision is made to return to shore.

Back aboard the *Beagle*, winter ice encrusts the skylight of Darwin's poop deck cabin. After a layover in Port Famine and some skirmishing with native people, the *Beagle* sails southward through the Magdalena Passage with Mount Sarmiento standing above, glaciers cascading to the sea in splendid beauty. The glaciers "seem doomed to last as long as this world holds together," writes Darwin.[56] (As climate warms, his conclusion seems less certain.)

Turning westward through the Cockburn Channel, the ship emerges into the Pacific, flanked by the aptly named East and West Furies, "a rock-studded constellation of small islets that posed a death trap for ships," with the Tower Rocks looming ahead.[57] Watching the many swells breaking in foam over the rocks, Captain FitzRoy names these treacherous waters the Milky Way. Darwin records, "The sight of such a coast is enough to make a landsman dream for a week about death, peril, & shipwreck."[58] Illness overcomes Darwin's friend George Rowlett, the ship's purser, whose burial at sea deepens the sense of peril.

Gales have blocked passage northward to the intended destination of Coquimbo. By June 28, 1834, they have reached the port of San Carlos (Ancud) on the Chilean island of Chiloé, where, Darwin notes, "submarine beds have been elevated into dry land only recently."[59] When the rain breaks, the fireworks spewed by Mount Osorno become visible to the northeast.

Darwin finds the scenery as sublime as that of Brazil. Bamboo vines climb through the trees; ferns abound. At the *Beagle*'s subsequent anchorage in Valparaíso, Darwin lodges with a Shrewsbury boarding school buddy, Richard Corfield, and from there undertakes excursions inland, continuing to find evidence of recent uplifts: seashells at 1,300 feet elevation, for instance. Upon visiting a copper mine, he finds men carrying water from the mine in skins on their backs—work that ought to be done by pumps. Poverty is widespread.

Following a bite from a vinchuca bug (carrier of Chagas disease, caused by a trypanosome organism), Darwin falls very ill. Fleas feast on him as well. A stunning volcanic eruption of Mount Osorno "spouting out volumes of smoke" greets his recovery in November; "jets of steam or white smoke" spew forth from another crater with a saddle-shaped summit.[60]

Months of making nautical charts take the *Beagle* to the southern end of Chonos Archipelago by Christmas Day 1834, which turns out to be "not such a merry one" as the men enjoyed "slinging the monkey" at Port Desire the year before. A waving shirt gets the crew's attention and a boat is sent to investigate. They find American sailors marooned for fifteen months, deserters from a whaler. This chance encounter with the *Beagle*, writes Darwin, prevented them from wandering "till they had become old men."[61]

During the night of January 19, 1835, the crew again watches Osorno as, "in the midst of the great red glare of light, dark objects in a constant succession" erupt and arc downwards. The red glare lights up the surface of the sea; by morning the volcano has "regained its composure."[62]

The next month brings even more spectacular geological rumblings as a great subduction zone earthquake centered on Concepción strikes the coast of Chile. Multiple quakes of great magnitude happen almost every century in Chile; when the crust ruptured this time, Darwin was in Valdivia, two hundred miles to the south, lying in the woods to rest:

> It came on suddenly & lasted two minutes (but appeared much longer). The rocking was most sensible; the undulation appeared both to me & my servant to travel from due East. There was no difficulty in standing upright; but the motion made me giddy. . . .
>
> An earthquake like this at once destroys the oldest associations; the world, the very emblem of all that is solid, moves beneath our feet like a crust over a fluid; one second of time conveys to the mind a strange idea of insecurity, which hours of reflection would never create.

Darwin continues to describe damage in the town and the "horror pictured in the faces of all the inhabitants." An old woman tells him that water rushed in "like an ordinary tide only a good deal quicker."[63]

Farther north the tsunami wave devastates the coastline and has "almost washed away" the wrecked homes of Talcuhano, the port city of Concepción. The quake destroys seventy villages. "The whole coast was strewed over with timber & furniture as if a thousand great ships had been wrecked."[64] Slabs of marine rocks from deep water lie cast upon the beach.

Talcuhano and Concepción present to Darwin "the most awful yet interesting spectacle" he has ever beheld. Witnessing the ruins of Concepción becomes the

Figure 3.3 Remains of the cathedral at Concepción, 1835, drawn by Lieutenant John Clement Wickham.

third of the three most interesting things Darwin experiences during the voyage (the other two being tropical vegetation and the Fuegians). How fortunate, he notes, that the quake has happened during the day: "In a large boarding school, the beds were buried 8 feet beneath bricks, yet all the young ladies escaped. The thatched roofs fell over the fires, & flames burst forth in all parts; hundreds knew themselves ruined & few had the means of procuring food for the day.—Can a more miserable & fearful scene be imagined?" Within the town of Talcuhano rests a schooner, deposited by a swift tsunami wave at least twenty-three feet high. The thought of a catastrophe of such magnitude leveling London distresses him: "England would become bankrupt."[65]

From March through June 1835, Darwin rides over the Andes to Mendoza and back again via the Portillo and Uspallata passes, where thousands of feet above sea level he observes beds of ancient seashells atop Andean ridges. The squalid conditions experienced by Chilean miners appall him as much as the genocidal wars of colonization.

He finds silicified trees, similar to living coastal species, entombed in marine sandstone, now resting seven hundred miles inland. His experience of Concepción's great earthquake together with these extraordinary transformations of landscape instruct him that Nature tosses blocks of the earth's crust up and down in deep geologic time with the same fury that bedeviled the *Beagle* rounding Cape Horn. The landscape is a slow-motion sea of geologic graybeards, swelling and crashing through vast time.

He rides north from Valparaíso to Copiapó at the southern end of the Atacama Desert, then sails to Lima in August to rendezvous with the *Beagle*, his terrestrial South American expeditions finally over, his fantasies of adventure in distant countries fulfilled.

Tortoise Pets and Children's Playthings

October 20, 1835, Tahiti; December 3, New Zealand. Darwin notes friendly Tahitians, unfriendly Maoris, and observes Christmas Day 1835 at the Waimate mission in New Zealand. By March 1836 he is in Australia, being introduced to its Aboriginal people. From there he sails to South Africa, where he meets John Herschel, the famed astronomer and philosopher of the inductive method, who is in Cape Town to observe Halley's Comet.

Crossing the Pacific and visiting the Cocos (Keeling) Islands, Darwin infers the origin of atolls. He hypothesizes that a marine volcano erupts underwater, and then eventually rises to the surface of the ocean. A fringing reef forms; after volcanism ceases, ongoing subsidence leads to the formation of a lagoon. An atoll is left as the result. Stephen J. Gould later refers to Darwin's reasoning as a triumph of intellect. Darwin has substituted place for time and thus arranged island variation in a temporal series based on a Lyellian causal mechanism: rising and subsiding.[66] One island is an example of another's future, a different island an example of its past. Darwin's atoll theorizing secures his position as a pioneering geologist and leading disciple of Charles Lyell's "principle of uniformitarianism": the present is the key to the past.

A Galápagos tortoise youngster accompanies Darwin back to England as a pet. Upon returning to England, he realizes that the tortoise dinners he ate at sea depleted the pool of specimens needed to test the hypothesis of island-by-island variation. He believes that while in the Galápagos, mariners introduced the

Figure 3.4 Pirate Darwin commandeers a Galapagos tortoise. Illustration by Jan Glenn.

 tortoises as a food source. As a result, island-to-island variation should reflect no more than accidental introduction.

Retroactively, he concludes that island-by-island variation of native Galápagos tortoises supports the case for life's mutability and descent with modification. Species vary from island to island but resemble those from nearby continental shores. Iguana species diverge to feed on land or sea. He realizes that variation among tortoises and finches provides evidence for his claims. The beaks of "Darwin's finches" eventually reign emblematic of this insight, a defining image for the origin of origins stories.

A brief stop at Saint Helena Island in the South Atlantic offers the chance to visit Napoleon's grave. On October 2, 1836, after a dash across the Atlantic to the coast of Brazil to recheck measurements of longitude and then back in time to seize a glimpse of a steaming volcanic crater in the Azores, the *Beagle* reaches anchor at Falmouth, England. On October 28 the *Beagle* docks in Greenwich. The chronometer reading by FitzRoy astonishingly differs from Greenwich time by only five seconds. Darwin has arrived home early for Christmas 1836, happily anticipating reuniting with his family.

Bedtime Stories

Darwin recorded in his *Beagle* journal stories of graybeard waves breaking over two-hundred-foot cliffs, of the cruelty of human slavery, of the poignancy of lost love, of skirmishes with native people able to sling rocks with deadly accuracy, of rides with gaucho soldiers across a war-torn frontier, of squalid living and foolish missionaries, and of the devastation wreaked by earthquake, volcano, and tsunami. Despite his complicity in the conduct of empire, Darwin never excused cruelty nor hesitated to condemn the actions of wicked men. His reflections spoke to the common bonds among disparate peoples, whatever extremes of difference appeared to separate them.

Throughout his rite of passage he collected with fervor and observed in astonishing detail the range of life from minute crustaceans to gargantuan mammals. He contemplated both the living and the dead and wove their lives together in grand drama.

Darwin cherished his souvenirs and memories. Bolas, stirrups and spurs, bone arrowheads, and boomerang—artifacts from the pampas to the outback—became his own children's playthings.[67]

Imagine papa Darwin enchanting his children and grandchildren at bedtime with stories of Philos's adventure and all of its dangers not "nearly so bad as he beforehand imagined." Then again, who could have imagined all of the great perils twenty-something Darwin endured?

4

IRRITATING WORMS

The Elderly Darwin Fascinated by the Intelligence of Worms

My older sister thinks she's so pretty.
I told her that no matter how much time she spends
looking in the mirror, her face will always look
just like her rear end.

Doreen Cronin, *Diary of a Worm*

When you and your older sister are both earthworms, there's a great deal of truth to this remark. To a younger brother worm, it's quite funny. To a Charles Darwin, it's a curious and interesting fact.

Worms typically pay little attention to mirrors, except, perhaps, when occasionally crawling across one. Indeed, a worm's head and tail ends are almost mirror images. It aids their motility. Being a tube with front and rear openings, and lacking any real head, is largely what makes a worm a worm. A worm's lack of a skeleton also enhances twisting and stretching. Body symmetry makes moving about in a world of excavated tunnels easier. And worms do tunnel. Much of the soil throughout the world has traversed worm guts to emerge as castings above burrow openings. Lowly worms do mighty work.

Man Is But a Worm

Worms are primitive organisms, yet they have mastered the basics depended on by more complexly limbed and skeletonized creatures. They crawl, dig, breathe, eat, excrete, and reproduce. Some swim. Earthworm bodies stack segments like so many muscularized, fluid-filled inner tubes. Some segments bristle. A worm snout has a small protrusion in front of the mouth to aid in digging; the mouth may pucker, forming a minuscule suction cup to grab debris and drag it to a burrow. Their skin, in responding to touch and other stimuli, is irritable. (And Darwin irritated worms on purpose.) Skin secretions help slide the worm across the ground, and gut discharges help dirt glide through the body. They crawl backwards as easily as forwards.

A little boy worm, however, cannot tease his little girl worm sister without making fun of himself/herself. Earthworms are hermaphrodites after all: each individual has male and female sex organs. Earthworm segments differentiate along an axis from front to rear end. Still, one segment does look pretty much like another, the ones up front more sensitive to light and all capable of sensing touch. Female pores in segment fifteen release eggs; sperm cells emerges from paired pores in segment sixteen.[1] Nerves, guts, muscles, skin, bristles, glands, pores: worm bodies are models of body development. On close inspection, worms are quite complex as well as elegant creatures. And the rear end does look pretty much like the front end, as generations of introductory biology students have noticed.

Thanks to a "strong taste for angling," Darwin encountered worms at a young age. Worms prompted his first sense of antipathy toward cruelty. As a means of being humane, he learned to euthanize the worms in concentrated saltwater before spearing them with a fishhook.[2]

Worms certainly intrigued the elderly Charles Darwin as well. He determined that his backyard earthworms paid no attention to the "deepest and loudest tones of a bassoon" played by his son Francis. "Shrill notes from a metal whistle" also interested Darwin's worms not at all. Nor did wife Emma's piano music or shouts—that is, unless the shouts were so close that the worm could feel the person's breath. Or the vibrations of the piano strings. Worms on a table ignored piano music from across the room; Mozart didn't faze them.

Worms that Darwin's research assistants (his sons) had placed in a pot of soil *on* the piano, however, at once crawled into their burrows when Emma played C on the bass clef. When she struck G above the treble clef, they similarly scurried into their burrows.[3]

Darwin felt that this response made good sense from a worm's perspective. Worms sense vibrations transmitted through solid material. Certain frequencies may alert them to the digging of a predatory mole—perhaps the pitch range between bass clef C and treble clef G.[4]

Always seeking precision in his observations, Darwin made worm science the family business. He put his son Horace to work looking for worm castings in the cellar of a neighbor's house. They were abundant in a corner of the structure that had been sinking. Francis set about collecting decomposing leaf fragments from worm burrows to test them for alkalinity.[5] William, a decade older than his male siblings, was assigned the task of recording whether worms dragged leaves into their burrows by the stalk or leaf tip. Two other brothers, Leonard and George, and older sister Henrietta (by this time married) wormed their way out of taking part in their father's slimy work.

From Root Tips to Worm Snouts

How curious that nearing the completion of his researches, dutifully assisted by Francis, his botany-destined third son, evolution's most respected sage spent his final days turning from roots and shoots to worm-watching. His study *The Movement and Habits of Climbing Plants*, published in 1875, detailed how plants, by virtue of vines and tendrils, were able to move. By 1880 he was ready to push further. In *The Power of Movement in Plants*, Darwin described the spiral traced by the growth tip of a stem and reported on experiments in root tip growth. Root tips, undisturbed, grow downward, in response, it was presumed, to their own weight.

Darwin experimented with cutting off root tips. Thus desensitized, roots grew sideways, not down. With no knowledge of hormones and tropisms (automatic, chemically mediated responses to physical stimuli by plants), Darwin reasoned that root tip response was analogous to animal instinct. He even presumed that a growing tip might "be compared to the brain of a simple organism."[6] For the elder evolutionist, the leap of science between root tip smarts and intelligent use of worm snouts was not large.

In 1881 Darwin, now seventy-two, published his final book, *The Formation of Vegetable Mould Through the Action of Worms with Observations on Their Habits*. By attending to the tip of the root and the snout of the worm, he observed how organisms, both plant and animal, probed and moved about, reacting to their surroundings. Curious indeed.

His worm and "vegetable mould" (humus) research was epic in scope, dealing with the fate of ancient ruins and the rudiments of mind. The habitability of the globe and adaptability of its creatures took center stage. He detailed how minute effects accumulated through vast time to accomplish mighty results, recapitulating the primary thesis of his life's work.[7] He argued that small changes, operating persistently, had transformed the earth on a grand scale. As evidence he detailed the production of "vegetable mould" (think of topsoil and the decomposing surface layer below the leaf litter) by common earthworms.

These features of the tips of organisms demonstrated remarkable responsiveness to the environment. They captivated Darwin. What was the aging Darwin still searching for? His studies evoked a child's enthusiasm for backyard nature, for poking about in the garden, under rotting logs, and along stream banks. He had wondered about organs of motility and sensitivity during many stages of his life, from teenager to adult. He believed that small things could reveal very basic aspects of biology—especially the antecedents of higher function. Did he wonder as well how to unite plants and animals to a common ancestor, a sensing, motile, adaptable fiber of primordial life?

With worms, he was especially curious about the emergence of intelligence—an ability to respond appropriately to the environment on the basis of past experience.

 "When their attention is engaged," he mused, "they neglect impressions to which they would otherwise have attended; and attention indicates the presence of a mind of some kind."[8]

On the Intelligence of Worms

Perhaps Darwin's worms practiced an elemental form of trial-and-error learning. With respect to drawing leaf litter objects into their burrows, he recorded "that worms try all methods until they at last succeed."[9]

Darwin even wondered about mental representations in the consciousness of worms. Using small triangles of paper as surrogate leaves, Darwin observed that worms drag the paper clippings into the burrow apex first. This behavior led to a stunning conclusion:

> If worms are able to judge, either before drawing or after having drawn an object close to the mouths of their burrows, how best to drag it in, they must acquire some notion of its general shape. This they probably acquire by touching it in many places with the anterior extremity of their bodies, which serves as a tactile organ.[10]

Darwin's worms could see with their snouts, by touch. His worms read leaf litter in Braille.

On April 15, Doreen Cronin's little worm enters in his diary: "I forgot my lunch today. I got so hungry that I ate my homework."[11] Harry Bliss, illustrator of *Diary of a Worm*, clearly depicts the little worm eating his homework apex first. Darwin did not comment on whether and how his worms ate their homework. On another occasion, though, Darwin experimented with pine needles. Pine needles come in bundles attached at the base—in this experiment, bundles of two. Mentally sharp worms pulled needle bundles into their burrow by the base and thus avoided snagging a pine needle point. Temperature, as it turned out, affected his worms' pine-needle-hauling industriousness:

> These worms, however, worked in a careless or slovenly manner; for the leaves were often drawn in to only a small depth; sometimes they were merely heaped over the mouths of the burrows, and so sometimes none were drawn in. I believe that this carelessness may be accounted for by the air of the room being warm, and the worms consequently not being anxious to plug up their holes effectually. Pots tenanted by worms and covered with a net which allowed the entrance of cold air, were left out of doors for several nights, and now 72 leaves were all properly drawn in by their bases.[12]

In the case of pine-needle hauling, Darwin concluded that the base of the needle held a special attraction for his earthworms. He tried gluing points together with shellac (then waiting for all the aromatic compounds to evaporate). He tied the paired points of the pine needles together with fine thread. Nearly always, whether warm and slovenly or cool and industrious, the worms pulled needles into their burrows by the base. At the conclusion of the pine needle trials, worms went on to prove their mettle by dragging the petioles of *Clematis montana* flowers into their burrows. Darwin clipped these flowers from the vines growing over his veranda.

Long before sacrificing veranda flowers to test earthworm IQ, Darwin had pondered how humans became human, how the social instincts of animals subjected to natural selection became the virtues necessary to civil society. From rudiments of mind in lowly and ancient creatures descended civilized behavior. No wonder a fascination with backyard worms framed Darwin's intellectual journey from child collector to distinguished researcher. Creation was not a puny six-day event but an everlasting, always inventive becoming. Life evolves; compassion emerges. A continuous creative birthing of novel forms, ever more remarkable than their primitive ancestors, proceeds through geologic time.

Figure 4.1 "Man Is But a Worm" from the cover of *Punch's Almanack for 1882*, drawn by Linley Sambourne and shaded by Paula Mikkelsen. Reproduced with permission from Paleontological Research Institution.

Following publication of *The Formation of Vegetable Mould Through the Action of Worms*, and five months before Darwin's death, *Punch's Almanack* published a cartoon depicting "Time's Meter" as a parody of the sage's latest intellectual diggings. *Punch* was forever pulling on Darwin's beard. In this cartoon a balding, aging, whiskered evolutionist in a Michelangelo pose watches as a march of monkey-like creatures unwinds. From one creature to the next, tails are shed and postures improve. The progression terminates in a top-hatted English gentleman, and the line begins with a segmented earthworm. The cartoon reinforces how little boys and girls descend, in essence, from lowly worms, perhaps prone to teasing each other as an expression of affection.

The pithy caption reads "Man is but a worm."[13] Ironically, points out Janet Browne, Darwin's preeminent biographer, the illustration mirrors the cycle of life and death. Burial returns all creatures to the earth where worms may feed on them.[14]

Flustra and the Firth of Forth

Publication of the treatise on worms preceded Darwin's death by a year. His delight in earthy creatures spanned a lifetime. In adolescence a particular group of invertebrate forms drew his attention. They demonstrated motility (at least as larvae) and irritability, the same as his backyard worms. He pondered these "polyps" as a young man. They planted in his mind the problem of common origins deeply buried in the history of life. They framed his theorizing in the final pages of *On the Origin of Species* where he pondered the ancient existence of an organism neither plant nor animal yet capable of becoming either.

His worm science, in effect, returned him to his backyard roots, equipped with a fully formed intellect and a seemingly inexhaustible reservoir of natural history knowledge. Blessed with an early childhood education of free-roaming the English countryside, and stimulated by working side by side with his brother in a homemade chemistry lab, Darwin found boarding school and medical college worse than dull. They required escape. Yet a few events transpired during his aborted foray into medicine that proved invaluable to his *Beagle* success.

During his two years in Edinburgh, Darwin accompanied his mentor, Robert Grant, on expeditions to collect sea pens and sponges from the Firth of Forth. Grant was a follower of Jean-Baptiste Lamarck and adhered to Étienne Geoffroy Saint-Hilaire's doctrine of transmutation of species through natural causes. "Transmutation" shocked God-fearing people because it ascribed to the living world the potential to mutate into new forms. Such material or natural cause offered explanation without recourse to independent creations or divine interventions.

Lamarck ascribed an inner drive to perfection as a primary reason for the transmutation of species. This inner drive worked its magic to transform simple worms into complex higher forms of life—new traits being acquired during a lifetime of striving, then inherited by the next generation. The Lamarckian moral is simply this: if you are a worm, strive to crawl faster and you may someday have great-great- . . . grandchildren with legs.

Long hikes with Grant thoroughly immersed the teenaged Darwin in this revolutionary ideology. According to his biographers Adrian Desmond and James Moore, while Darwin was a student,

> the few evolutionists kept their heads down. . . . If nature and culture were *self*-evolving, if the clergy could not point to miraculously created species as a sign of His power operating from above, the Church's legitimacy was undermined. The logic was stark—even if it was rarely spelled out. The day people accepted that nature and society evolved unaided, the Church would crash, the moral fabric of society would be torn apart, and civilized man would return to savagery.[15]

Throughout his life Darwin walked in the company of those who foresaw rupture of the moral fabric and collapse of civil society if belief in transmutation of species became widely held. Yet he dared tread that path, and its initial steps required close examination of sea sponges and moss animalcules (bryozoans). Not worms, but still rather lowly, and with patterns of growth neither fully plant nor animal in the understanding of nineteenth-century naturalists.

Sponges and their simplicity were crucial to Grant's evolutionary theorizing; Lamarck's too. These men believed in environment and climate as agents of species change, not "a series of divine Creations."[16] At Grant's side, Darwin visited Newhaven fishers to examine their by-catch of sea slugs, sea pens, sponges, polyps, eggs, and larvae. Among them was *Flustra*, a colonial organism that grows as an encrusting mat with fronds centimeters in length. (Now flustrids are recognized as members of the Bryozoa, or "moss-like animals.")

Darwin's library on board the *Beagle* muddled the relationship between coral-building polyps (true corals, the cnidarians of today), encrusting mats (bryozoans), and coralline algae (plants)—in part because the biodiversity of marine organisms was poorly understood and in part because Lamarck and other pre-Darwinian transmutationists reveled in the search for primordial zoophytes. The term "zoophyte," proposed by Linnaeus, actually meant animal-plant intermediates, a reasonable idea. Experiments with the freshwater polyp *Hydra* had engrossed naturalists throughout the eighteenth century. *Hydras* moved like animals and regenerated like pruned plants; were they tentacled or branched?[17] The creature greatly confounded the notion of a plant-animal boundary.

Thus Darwin worked with an inadequate system of classification which he subsequently improved upon by concluding that his coralline specimens *Amphiroa* and *Corallina*, for example, belonged to the algae.[18] He drew marine creatures with great care and observed that in *Clytia* and *Flustra* (bryozoan polyps), a ring of ciliated tentacles surrounded the mouth. Encased in carbonate shells, all of these organisms grew in a branching pattern.

Darwin charted unknown anatomical territory first at Edinburgh and later from his *Beagle*-bound laboratory.[19] His close observations revealed a digestive tract, a discovery crucial to distinguishing polyps from plants. Gut development further divides the polyps into distinct groups. Bryozoans, for example, have both a mouth and an anus; true jellyfish medusae and coral polyps have just one dual-purpose opening.

The colonial organisms that captivated Darwin attach themselves to surfaces such as shells, rocks, and kelp. His *Zoology Notes* are replete with their technical descriptions.[20] Comparisons among their branching patterns of growth, organs of motility, rings of ciliated tentacles, as well as other structures at minute scales—jaw-like appendages, for example—fascinated him.[21]

For Grant, colonial sponges and encrusting bryozoans (known to his contemporaries as "sea-mats") suggested a common origin of plants and animals through transmutation from an undifferentiated zoophyte. Coral polyps and algae, especially their free-swimming gametes, likely held clues. At the meeting of the Wernerian Society in 1826, his student Charles Darwin (just seventeen years old) reported the discovery of the eggs of the wormlike parasitic skate leech, *Pontobdella muricata*, adhering to oyster shells.[22] Skate leeches hatch from an ovum within a capsule at the end of a ridged and fibrous-looking stalk resembling seaweed. Adult leeches suck blood from bottom-dwelling flatfish. Darwin's careful observations revealed the development of the leech embryo within the gelatinous egg, demonstrating that what was thought to be a young seaweed (mistakenly identified as *Fucus loreus*) was in fact an animal egg case.[23] Finding that tiny worm hidden within a stalked capsule was his first scientific breakthrough.

Shortly afterwards, on March 27, 1827, Darwin presented a scientific paper to the Plinian Society that described the free-swimming larvae of *Flustra*. Cilia propelled these larvae through the water; Darwin demonstrated that these were not motile eggs but hatchlings. Given *Flustra*'s uncertain affinity to animals, its filamentous system of propulsion was quite interesting to him and his mentor, Grant. Seaweed-like appearances disguised animal creatures in both instances, *Pontobdella* and *Flustra*, leech and coralline. Wiggly motility—filamentous sinuosity—characterized them both. Grant had little doubt that these observations pointed toward a common origin for both plants and animals.

From his teenage years onward, Darwin never overlooked opportunities to scrutinize the details of the corallines. *Flustra* caught Darwin's attention first at the Firth of Forth. Several years later off the coast of Patagonia he would describe a planktonic organism named *Cellapora eatonensis*, a relative of *Flustra*. He noted another species of *Cellapora* amidst other bryozoans netted near the Galápagos Islands and collected flustrids from the kelp beds of East Falkland Island and from the channels dissecting the islands of Tierra del Fuego.

The final few pages of *On the Origin of Species* returns to the problematical existence of a primordial zoophyte, ancestor to all plants and animals. Darwin very early surmised an analogy between coralline polyps and "turf-forming plants," entering in his *Beagle* voyage zoology notes, "I think there is much analogy between Zoophites & Plants, the Polypi being buds."[24]

Darwin was never quite so convinced as Grant that he had in hand proof of the unified ancestry of animals and plants. The anatomy of *Flustra*'s curious "beaks"—very un-plantlike structures—impressed him greatly. In fact, the structure and physiology of both plant and animal filaments would hold his attention for his entire life. In 1875, for example, his treatise on insectivorous plants discussed at length filament response to irritation and speculated that physical disruptions (e.g., folding) could disturb electrical currents within plant leaves.[25] The hair-like filaments and collapsing leaves of sundews and Venus flytraps meant doom for small insects. Plants that preyed moved rather quickly—perhaps by utilizing mechanisms shared with animals. Even so, the case for the common origin of plants and animals remained inconclusive in 1859. No organism, fossil or living, then stood as a candidate for such a role, worm or otherwise. Yet the shadow of an ancestor—his literal ancestor—haunted Darwin's mind.

Grandfather's Filament

Darwin's zeal for filamentous structures ultimately led him to reconsider grandfather Erasmus Darwin's musings on the existence of a primordial living filament or "embryon fibre," as celebrated in Dewhurst Billsborrow's 1794 poem.

> How the first embryon fibre, sphere, or cube,
> Lives in new forms,—a line,—a ring,—a tube.[26]

The phrase "the first embryon-fibre" antedates both cell and Darwinian theory, yet it anticipates the evolution of body forms. Billsborrow's poem celebrates the spiritual romance of embryonic, primordial life climbing from brute to angel; of irritable, excitable filamentous structures becoming the swimming, crawling, running, flying creatures of land and sea. He was, as was Lamarck, inspired by Erasmus Darwin's contention that all animals "have a similar cause of their organization, originating from a single living filament, endued indeed with different kinds of irritabilities and sensibilities."[27] Similarity of form impressed the transmutationists. Erasmus explained:

> When we revolve in our minds the great similarity of structure which obtains in all the warm blooded animals, as well quadrupeds, birds, and amphibious animals, as in mankind; from the mouse and bat to the elephant and whale; one is led to conclude, that they have alike been produced from a similar living filament. In some this filament in its advance to maturity has acquired hands and fingers. . . . In others it has acquired claws or talons, as in tygers and eagles. In others, toes with an intervening web, or membrane, as in seals and geese. In others it has acquired cloven hoofs, as in cows and swine; and whole hoofs in others, as in the horse. While in the bird kind this original living filament has put forth wings instead of arms and legs, and feathers instead of hair. . . . And all this exactly as is daily seen in the transmutations of the tadpole, which acquires legs and lungs, when he wants them; and loses his tail, when it is no longer of service to him.[28]

Clearly, the younger Darwin was not the first to wonder about the mystery of mysteries (life's origins), nor the first generation within his family to speculate on the origins of species. In succinct terms, grandfather Erasmus asked, "Shall we conjecture that one and the same kind of living filaments is and has been the cause of all organic life?[29] Grandfather answered with a "yes," of course.

What seems of particular interest in the teenager Charles's first scientific paper is its devotion to filaments of motility. Juxtaposed with his interests in tendrils, vines, root tips, and worm snouts in old age, this beginning acquires a lifetime of significance. Attention to the "embryon-fibre" was present at the creation of his intellectual identity and evolved with it. Curiously, an essay complementing his grandfather's biography was his final piece of published writing.

Mixed up in Erasmus Darwin's, Saint-Hilaire's, and Lamarck's nascent evolutionary theorizing was a metaphysical concept of inner drive to perfection leading to improvement during a lifetime of striving. Life reaches upwards and so progresses. Analogously, seeking God improves human character and increases virtue. Billsborrow recognized this parallel between spiritual growth and transmutation of species. Material causes bring forth an angelic destiny. Heretical to Anglican orthodoxy Erasmus's contentions may have been, but in 1800 they reconciled science with mysticism. For the poet Billsborrow, Erasmus Darwin's philosophizing heightened his sense of religiosity. Transmutation united the material and spiritual realms.

The mature Charles Darwin eschewed any muddle-headed mysticism in his approach to science and demanded to know just how and in what ways life transformed itself, introducing novelty and complexity along the way. Divine guidance and divine intervention would not do. For him, movement indicated embryonic volition, and volition set life apart from inanimate matter. Filaments moved—and therefore contained clues about deep origins. Embryons—or embryos—developed, as did tadpoles in becoming frogs.

The potential to transform held by embryos and the ability to respond to the environment displayed by filaments must have puzzled Darwin greatly. They signified the origins of novelty and volition. The tiny skate leech ensconced within a seaweed-looking egg was both filamentous and embryonic. As a notable first discovery—Darwin's entry point in the scientific literature—it may have exercised an archetypical influence over his subsequent strivings to create a theory of the origin of species.

Simple forms, excitable and responsive to the environment, likely held in his mind the key to how life began, to what made life evolvable in the first place. Why was Grandpa's hypothesized first "embryon-fibre" capable of becoming a line, a ring, a tube—a hand, a wing, a claw? The modern answer stems from the properties of DNA strands, subjected to winnowing by the demands of survival. The existence of DNA strands, not to mention the mechanisms of inheritance, were unknown and unknowable in the nineteenth century, of course. Today's materialist account of life's origins claims that coded information, accumulating and mutating, is capable of chemically transforming inert matter into living organisms. Varying the code changes life, and interaction with the environment mutates the code.

These ideas emerged as Darwin and Darwinism matured—many years after the voyage of the *Beagle*. Two years of studying medicine at Edinburgh had persuaded Darwin to seek a career elsewhere. In part to please his father, he was bundled off to Christ's College in Cambridge, supposedly to enter the Anglican ministry. A humble rural parsonage beckoned, with ample time to engage in natural history. Later in life he reflected on his proclivities and interests which had turned out to suit him well: "I had strong & diversified tastes, much zeal for whatever interested me, & a keen pleasure in understanding any complex subject or thing."[30]

Becoming a naturalist was the chosen path for pursuing his "zeal for whatever interested" him. And it promised excitement—the kind of excitement shared by Alexander Humboldt in his romantic narratives of travel in the New World tropics, popular reading among Cambridge students.[31] Darwin's Cambridge education solidified his life's calling as a naturalist.

Clerics at Christ's College no doubt disapproved of Erasmus Darwin's musings, and they exerted a primary influence over Charles Darwin's mind following his Edinburgh experiences. From required reading of William Paley's *Natural Theology*, Darwin learned that all things were fitted to their circumstances by Divine Providence. The world was the product of a Designer. He who was without cause was the final cause of order in the universe. God decided whether an anus was needed, where to put it, whether or not to surround it with tentacles, and whether or not calcareous secretions would be useful. He left *Flustra* and its kin to fluster the evolutionist. Exquisite adaptation of living things to one another, and their circumstances produced harmony, happiness, and goodness in keeping with the Designer's intentions. Heartfelt grasp of these intentions was the goal of faith and the reason to study nature. Such was the instruction Darwin received in preparation to be not just a naturalist but also a pastor.

Darwin had heard Paley's Designer views ridiculed in Edinburgh by Grant and his followers. These academics postulated that "dead atoms" and animal tissue, according to natural cause, of and by themselves could bring forth complex order and improvement. Their atheism was at odds with notions of "intelligent designing minds."[32] Tested on Paley at Cambridge and instructed by Grant at Edinburgh, Darwin packed the conflicting teachings of design and transmutation into separate bags as he boarded the *Beagle*. He had no real appreciation yet for how the journey would transform him intellectually and set him on a path critical of both Lamarckism and creationism.

Budding Branches

From fourteen fathoms down off the coast of Patagonia, Darwin scooped up the coralline *Cellepora*. He found *Flustra* in the kelp beds of the Falkland Islands, believing it to be related to *Cellepora*. More corallines appeared in the waters surrounding Tierra del Fuego and the Chonos Archipelago of Chile. With these specimens in hand, Darwin engaged himself in an "orgy of comparative anatomy" as he focused on rings of tentacles, mouths, and anuses in relation to polyps, stalks, and capsules. His work contributed to the understanding of how animals had adopted sessile lifestyles in marine habitats, waiting for food to come to them. Ultimately, Darwin's careful descriptive work foreshadowed future clarification of bryozoan biology.[33]

The search at microscopic scale often found inexplicable novelty, not simplicity. For example, the flustrid-like *Cellepora*, a "moss-animal" echoing forms from the Firth of Forth that had fascinated Darwin in Scotland, sported tiny structures resembling a vulture's beak. There were seventy-five vulture beaks to the inch, each seated on a "peduncle" (thin stem) that could rotate as well as move out and in. Its movement responded to irritation and occurred rapidly; the function of this structure remains unknown, though a role in defense has been proposed.[34]

Be careful not to underestimate the power of the imagery of branching corallines that Darwin netted off the coasts of South America during the voyage of the *Beagle*. Keep in mind that they were sister species to Scotland's *Flustra*, featured in Darwin's first researches. Images of these organisms, collected, dissected, pickled, and curated, danced in Darwin's head. Over time, imagery guides thinking.

Everyone who studies Darwin becomes familiar with the single figure embedded in the *Origin of Species*: the branching diagram, the image he used to demonstrate the effects of natural selection.[35] The stature of this diagram has grown to become *The Tree of Life Web Project* (not to mention a delightfully illustrated children's book, *The Tree of Life* by Peter Sís).[36] From so simple an illustrative beginning, a remarkable online archive has, and is being, evolved. Darwin himself referred to his diagram as depicting the Tree of Life:

> As buds give rise to growth by fresh buds, and these, if vigorous, branch out and overtop on all sides many a feebler branch, so by generation I believe it has been with the great Tree of Life, which fills with its dead and broken branches the crust of the earth, and covers the surface with its ever branching and beautiful ramifications.[37]

Generations of students and scholars know the lingo and diagram of branching, the numerous arbor-rooted metaphors for life's history. It may be the most widely grasped representation of Darwinism. A German art historian, Horst Brederkamp, has proposed a startling origin for Darwin's branching tree. An even deeper image may have generated it: the growth pattern of the coralline *Amphiroa orbignyana*, a crusty form of sea life resembling *Flustra* but now classified as algae.[38] Brederkamp overlaid a specimen Darwin collected onto Darwin's famous diagram in the *Origin*.[39] The branchings aligned with startling consistency.

As Florence Maderspacher, an editor of *Current Biology*, tellingly notes, "Sometimes an image can stimulate us to think in a new, unanticipated direction." Guiding Darwin was perhaps not just the image "of the much quoted tree, but that of a coral."[40] Remember that corallines constituted a confusing set of nineteenth-century species. Members of the corallines are now understood to be either colonial animals or calcareous plants. Both commonly exhibit a branched pattern of growth, as do some forms of true corals. Darwin's "coral," as suggested by Brederkamp, however, is actually a calcareous red alga—neither coral nor bryozoan.

Soon after completing the *Beagle* voyage, Darwin duly recorded a coralline-inspired thought in his "Notebook B" on the topic of the transmutation of species: "The tree of life should perhaps be called the coral of life."[41] He had observed that new calcareous growth—whether with polyps or not—grew from the tips of branches. New growth symbolized new life; base branches suggested extinct species. The image, despite the confusion of nomenclature (and plant and animal affinities), worked.

In the notebook a sketch of this image quickly followed. Branching points indicated common ancestors. He explained that the most recent branchings placed species near to one another—buddings separated by "the finest gradation." The tips of branches far apart stood for an "immense gap of relation."[42]

Darwin's diagram depicted divergence of characters within genera of organisms through time due to natural selection. The branching points were hypothesized common ancestors. Near the trunk, as lineages converged upon a common

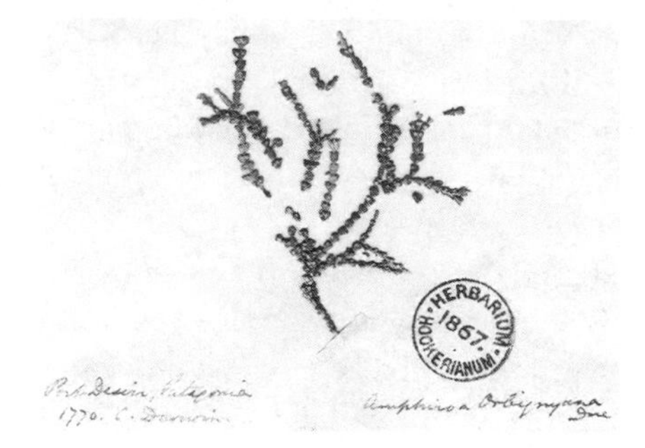

Figure 4.2 The branching coralline *Amphiroa orbygnyana*, a calcareous red algae, collected by Darwin while at Port Desire, December 1833. ©The Trustees of the Natural History Museum, London. Used with permission.

ancestor, branching points suggested what grandfather Erasmus had imagined: that a ring may have descended from a filament, a tube from a ring.

The case for coralline imagery functioning at the root of Darwin's thinking holds remarkable appeal. It links his Tree of Life to his earliest tutelage by Grant, who in turn found substance in Erasmus Darwin's poetic speculations. Like corallines, lineages branched and budded new populations, variation distancing them from one another and common ancestry uniting them.

Darwin's "coral" is not a coral but a red alga, ironically named after a rival naturalist, Alcide d'Orbygny, who journeyed in South America just prior to Darwin. Perhaps the nomenclature of evolution should be changed to "The Algae of Life."

Atoms of Life

Grant's speculating merged plants and animals in a single starting point. He claimed that algae were related to polyps and postulated continuity from the most primitive to the most complex.

Lamarck argued also that higher animals evolved from "simplest worms"—the line, the ring, the tube.[43] Lamarck's science and Erasmus Darwin's poetry had dwelt upon the capacity of a primordial filament to generate new forms well before the departure of the *Beagle*. Darwin's branching imagery descended, in part, from these influences. To escape their mystical overtones, he introduced natural selection acting on chance variations, but not without an intellectual struggle. During his formative years, "natural" also referred to theological accounts of nature's beneficent order. It was God's design. "Natural" and "divine" joined hands in Paley's theology of Providence.

Perhaps visions of Erasmus Darwin's primordial filament—from which subsequent, more complex forms of life presumably sprang—danced in Charles Darwin's head as he watched flustrid peduncles twitching and turning. Perhaps not. Complex, novel organs—peduncled vulture beaks—if derived from a primordial, undifferentiated filament, obliterated the imagined simple ancestral structure. Even so, the movements observed in micro-scale seemed significant. In 1834 Darwin again recorded observations of a moving "bristle" on a coralline. Whether these organisms were plants or animals, movement seemed to be due to sensation of the environment and coordination by a nervous system—what Darwin described as "co-sensation and co-will," fundamentals of life.[44]

Darwin accepted the notion that physical contact might disturb electrical currents within plants and cause stems, leaves, or tendrils to bend and fold in adaptive ways. In the flustrid corallines, he concluded that a relationship between sensation and will—his presumed basis of volition—suggested the existence of a nervous system and thus implied animal status for these organisms. In 1881 he argued very explicitly that, with regard to earthworms, volition influenced behavior and was a product of their primitive central nervous system. The imagery of irritable, sensible primordial filament, progenitor of plants and animals, lurked in these thoughts.

This sensitivity and irritability were properties that presumably made the "embryon-fibre" adaptable to its surroundings, thought Erasmus Darwin. So too did movement of both plants and animals in response to the environment hold Charles Darwin's attention. He puzzled over the phenomena of sensation and willful movement across very different organisms—vine tendrils, root tips, and worms at both ends. All these movements must have echoed the motility and responsiveness observed in the tiny structures and functions of flustrids.

Something planted in his mind at age eighteen prompted him to return again and again to the details of the coralline *Flustra*, an inauspicious organism populating waters from Scotland to the Falklands, from Patagonia to Tierra del Fuego. Among the features observed within the primitive corallines, *Flustra* preeminent among them, he likely imagined finding the residue of "the zero point, where the plants and animals meet."[45]

Collecting corallines occupied Darwin at many junctures during the voyage. According to Richard Keynes (Darwin's great-grandson and editor of his Beagle *Diary* and *Zoological Notes*), examining the horde of marine invertebrates did not shake his Designer leanings. Keynes asserts that "it is difficult to read into [Darwin's notes] views on the transmutation of species that he had not yet begun to develop seriously in any other context."[46] At the same time, Darwin's interest in the details and distribution of flustrids was striking, generating imagery and thought that took decades to crystallize.

Trailing the *Beagle*, Darwin's plankton net added to his store of marine critters of small scale that wriggled and jiggled and tickled to his lifelong delight—and inspiration. He sought to understand the macrocosm of life in the minuscule forms of the sea. These creatures were his first love. Their reproductive cells and larvae harkened back to the "ultimate atoms of life."[47]

Sandwiched between *Flustra* and his grandfather's biography, between Grant's teachings and the echoes of Paley's natural theology was Darwin's life's work on variation and selection, on barnacles and orchids, on island finches and fossil quadrupeds, on pigeons and sundews, on seeds dispersing and worms burrowing. Dreaming about such grandiose accomplishments did not preoccupy him while on the *Beagle*. *Flustra*, the moss-animal bryozoan, had unwittingly launched his career as a naturalist. A coralline algal organism imprinted his mind with the imagery of branching. Ultimately, he might have imagined that the twitchings of Great- x 10^{googol} Grandfather Worm resembled the motion of a flustrid peduncle. Perhaps he surmised that the machinery coordinating the movement of its larval cilia was similar to that controlling the sinuous swimming of a skate leech or worm.

On board the *Beagle*, however, there was still enough of the Paley in twenty-something Darwin to keep him from diving too deeply into the waters where Lamarck and Erasmus Darwin had swum. Careful descriptions of his prized flustrids—the encrusting corallines—satisfied his emerging ambitions as a naturalist, refining the talents he displayed at scientific meetings while still only eighteen. The budding and branching pattern of growth among Darwin's corallines would come to undermine his belief in design. He entered in his zoology notes of 1834 the prescient statement "I examined the Polypus of this very simple Flustra, so that I might erect at some future day, my imperfect notions concerning the organization of the whole family of Dr Grants Paper."[48]

His statement at the time referred to the restriction of corallines to plants and the careful elucidation of encrusting colonial polyps as animals. His "imperfect notions" did more than systematically separate plants from animals within the corallines. When later interpreted in light of island-by-island variation among Galápagos mockingbirds, finches, and tortoises and the distribution of fossil megabeasts across Patagonia, they bequeathed an understanding of how lineages diverge, best mapped as the Tree of Life. The small parts of small creatures played a key role, and so, in grandfatherhood, he turned to backyard worm-watching with help from his family, looking to perfect his imperfect notions with detailed knowledge of the lowly creature. Setting sail from his country home, his shovels replacing the *Beagle*'s nets, he trawled the dirt for worms.

As noted earlier, he became interested in "how far they acted consciously, and how much mental power they displayed." By experimenting, he learned that "they try in many different ways to draw in objects" into their burrows. Success was due neither to chance nor to a special instinct for handling each particular object. Darwin concluded that "worms, although standing low in the scale of organization, possess some degree of intelligence."[49] For example, worms "knew"

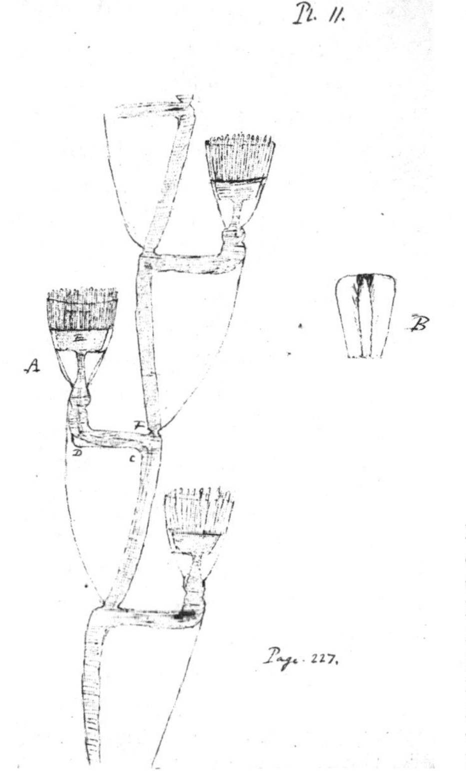

Figure 4.3 Darwin's sketch of a representative coralline, *Clytia*. "In the middle of the tentacula there is the mouth." The organism is a bryozoan. Richard D. Keynes, *Charles Darwin's Zoology Notes and Specimen Lists from H.M.S.* Beagle (Cambridge: Cambridge University Press, 2000), 202–3. Collected near East Falkland Island, March 1834. DAR 29.3: 61r. © Reproduced with the permission of the Cambridge University Library

to leave their burrows when the digging of a mole caused the ground to vibrate. The little worm boy in *Diary of a Worm* understands what such vibrations mean, too. He knows to dig deeper when he can feel the ground shake from people digging for bait during fishing season. No worm wishes to end up as fish bait or mole food. Still, many do.

Darwin found worms to be "timid." He believed they took pleasure in eating. An entry in *Diary of a Worm* makes the same point: "Never bother Daddy when he's eating the newspaper."[50] The little worm makes no mention of the intimate life of his parents, but Darwin determined that among worms, "sexual passion" was "strong enough to overcome for a time their dread of light." Apparently, turning down the lights helped to ignite their amorous nature, more so than bassoon or piano music. "They perhaps have a trace of social feeling, for they are not disturbed by crawling over each other's bodies, and they sometimes lie in contact."[51] While in contact, hermaphroditic earthworms exchange sperm with each other.

"It's not always easy being a worm. We're very small, and sometimes people forget that we're even here," writes the little worm in his diary. Darwin noticed. Inconspicuous forms of life mattered to him. He calculated that up to ten tons of dirt passed annually through the guts of earthworms in each acre of English countryside, "so that the whole superficial bed of vegetable mould passes through their bodies in the course of every few years." Only organisms "still more lowly organized, namely corals," concluded Darwin, had trumped this massive engineering feat "in having constructed innumerable reefs and islands in the great oceans."[52] From little worms and polyps mighty landscapes and seascapes grow.

Mentor Grant taught young Darwin to ponder the wormlike progenitor of both plants and animals. Darwin never solved this puzzle. In lowly forms he encountered surprising complexity: bristles, tentacles, "vulture beaks." Branching accounted for the diversity of lineages in terms of a shared ancestor, and he believed that this principle united even the plant and animal kingdoms. His American friend and colleague Professor Asa Gray of Harvard informed him that the reproductive bodies and spores of "lower" algae alternated between animal and vegetable characteristics. With respect to the "broad differences" between the plant and animal kingdoms, Gray claimed that they

> vanish one by one as we approach the lower confines of the two kingdoms, and that no absolute distinction whatever is now known between them. It is quite possible that the same organism may be both vegetable and animal, or may be first the one and then the other. If some organisms may be said to be at first vegetables and then animals, others, like the spores and other reproductive bodies of many of the lower Algæ, may equally claim to have first a characteristically animal, and then an unequivocally vegetable existence.[53]

Such an insight from a renowned botanist pleased Darwin. It echoed his efforts to tease apart and identify the globular egg cases of the skate leech, *Pontobdella muricata*, from an early stage in the development of the rocky intertidal brown alga *Fucus loreus*. It made sense in terms of the corallines encompassing what became known as both animals (the bryozoans such as *Flustra*) and plants (the calciferous green algae genus *Halimeda*).

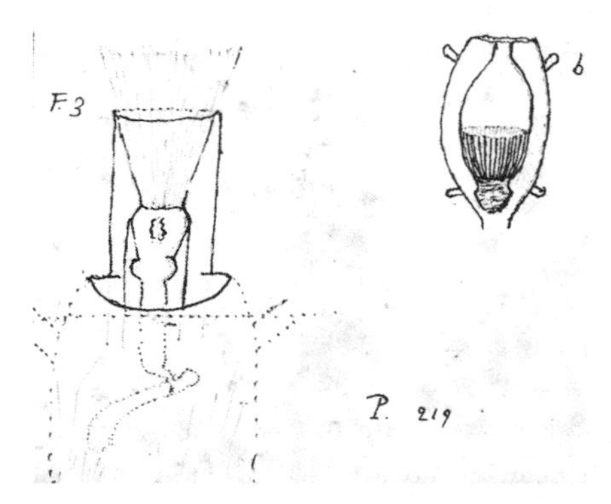

Figure 4.4 Darwin's sketch of *Flustra*; "the arms are enclosed in a transparent case; which is protrudable." Note how the ring of tentacles enclosing the mouth is extended on the left and retracted on the right. Richard D. Keynes, *Charles Darwin's Zoology Notes and Specimen Lists from H.M.S.* Beagle (Cambridge: Cambridge University Press, 2000), 194–95, plate 8, fig. 3. Collected in Tierra del Fuego, March 1834. DAR 29.3: 62. © Reproduced with the permission of the Cambridge University Library.

Primordial Forms

At the end of *On the Origin of Species*, Darwin concluded that descent with modification proceeded from "four or five progenitors" among animals and from "an equal or lesser number" among plants.[54] Modern biology posits six kingdoms of organisms, plus viruses, not just the stereotypical two of plant and animal. Nonetheless, Darwin's notion of a small number of discrete patterns of organization, from which all subsequent variations and complexities have emerged, was sound.

Only by analogy, admittedly a "deceitful guide," did the conclusion follow "that all animals and plants are descended from some one prototype." So far as was known to Darwin, "excepting some of the very lowest [organic beings], sexual production seems to be essentially similar. . . . [A]ll organisms start from a common origin."[55] Given the vast record of "divergence of character" from a common origin, Darwin concluded that the living of one epoch had descended with modification from those of an earlier time, winnowed by natural selection while randomly drifting to an unknown degree.

To admit to these conclusions meant admitting to the idea "that all the organic beings which have ever lived on this earth may be descended from some one primordial form."[56] For Darwin, this inference rested confidently upon analogy and extrapolation. Erasmus Darwin's primordial filament, the vital "embryon-fibre," remained an unknown entity, despite his grandson's lifelong attention to lowly,

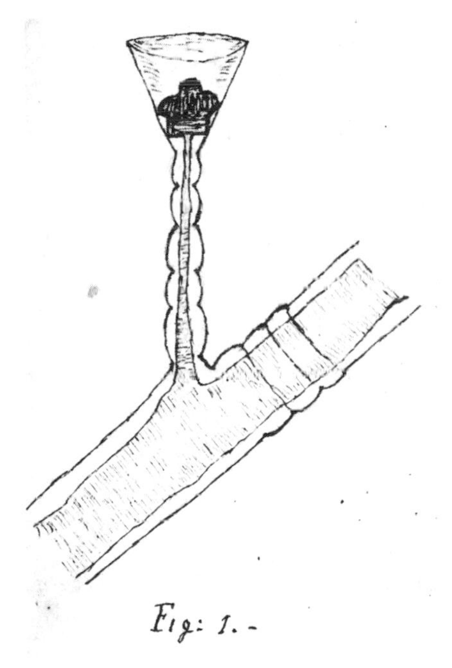

Figure 4.5 Darwin's sketch of "a polype retracted in its cup, with the peduncle rising at a joint in a branch," related to *Clytia*. Richard D. Keynes, *Charles Darwin's Zoology Notes and Specimen Lists from H.M.S.* Beagle (Cambridge: Cambridge University Press, 2000), 95, plate 7, fig. 1. Collected at Bahía Blanca, September 1832. DAR 29.3: 71r. © Reproduced with the permission of the Cambridge University Library.

 irritable forms and motile structures. Grant's conjecture stood unproven, though it lurked in the mind of his student for a lifetime, animating the imagery that guided his thought.

Twenty-first-century science tries to answer Grant's conjecture by following the telltale trail of DNA's similarity across organisms. *The Ancestor's Tale* by Richard Dawkins reveals this story, both its well-accepted branchings and its inconclusive analyses of relationship through descent among the earliest forms of life on earth. (Complex mechanisms of gene transfer and fusion cloud the picture.) Cellular life without a nucleus (prokaryotes), but still with DNA to translate into the machinery for living, constitutes the deepest root of the Tree of Life. The origins of eubacteria—*Escherichia coli* and company to greatly oversimplify—trace back to at least 3.8 billion years ago. Archaea compete for a truly ancient pedigree as well; some modern forms thrive in hot springs.

Cells with nuclei (eukaryotes) found themselves busy evolving by a couple of billions of years ago, and they probably got a boost from novel ways in which the non-nucleus clans teamed up with the nucleated ones. Hundreds of millions of years of experimenting yielded foraminifera, choanoflagellates, slime molds, ciliates, diatoms, amoeboids, euglenids, and a host of other distinct groups with even more esoteric and entertaining names. They indeed gained in complexity and diversity, but bodies—the triumphant structure of multi-celled organisms—took a while longer. By a half-billion years ago, complex bodies had begun to leave a rich fossil record.

Animals perch on a branch not too far from choanoflagellates and rather surprisingly (according to molecular-level analyses) close to fungi. Choanoflagellates propel themselves with a flagellum. As known to Robert Grant and Charles Darwin, sea sponges produced mobile "choanocyte" cells, each with a flagellum. Instead of swimming freely, sponge choanocytes undulate in harmony, creating currents that wash nutrients through sponge bodies. Nevertheless, the choanocyte-choanoflagellate similarity is unmistakable.

Grant and Darwin were on to something. Darwin's lifelong fascination with motility and the structures of minute forms of life—polyps and protists and gametes and worms—was well warranted, and their evolutionary relationships remain a challenging puzzle to solve. Land plants, however, find themselves on a branch very distinct from the one with fungi, sponges (and other animals, such as humans), and choanoflagellates. There seems to be no creature unmistakably part plant, part animal. Their ancestors were different kinds of organisms that had not become plant or animal in any sense Darwin would recognize.

How all forms of life ultimately unite in a common origin continues to tax evolutionary science.[57] The evidence of common ancestry resides in the degree of shared sequences of DNA in the double-helix strands that constitute life's primordial filament, much tinier than any flustering flagellum or peduncle seen moving under Darwin's microscope.

Sister Gets the Last Word

As Darwin noted, worms often find shelter beneath stones. They hollow out burrows and deposit their castings next to the stone. Bit by bit, worms excavate the ground from beneath a stone and the stone inevitably sinks. The process has yet to complete this effect at Stonehenge, though well it might.

In this sense, the work of worms expresses Darwin's conception of evolution metaphorically. Bit by bit, all things change. Life evolves in nearly imperceptible, and ultimately irresistible, steps. The relentless work of natural selection, winnowing through endless variants, transforms the world's biota.

At his big sister Caroline Wedgwood's estate, Leith Hill Place in Surrey, rested two large stones in a field where once a kiln had stood. One was sixty-four inches long, seventeen inches across, and ten inches thick. The other was a few inches longer, over three feet across, and fifteen inches thick. Darwin spoke with an elderly worker who remembered the two sandstone rocks resting thirty-five years before upon "a bare surface of broken bits and mortar," now turned to turf and decaying matter ("vegetable mould"). The smaller stone had sunken one and a half inches during that time; the larger one, closer to two inches. Accounting for rainwater and sheetwash removing the finest dirt excavated by worms, Darwin estimated that in two and a half centuries the stones would no longer protrude above the ground in Caroline's field.[58]

Some fifty years earlier they had corresponded through dozens of letters as he collected his *Beagle* treasures. Now Caroline's little brother Bobby had come to the end of his life's investigations, fussing about earthworms. He was enamored of lowly creatures akin to the skate leeches that had fascinated him as a teenager. Watching her brother putter around the rocks of her estate measuring quantities of worm castings, Caroline would have said, "Really, Charles, your interests long ago and today are as similar as the two ends of a worm." Caroline, as a last teasing, could have entered in her diary:

> My younger brother thinks he's so clever.
> I told him that no matter how much time he spends
> studying life's history, his findings will always sound
> just like his first discovery.

5

A LUNGFISH WALKED INTO THE ZOO

On the Origin of Limbs from Lobe-Fins

> *At the edge of the woods there was a pond, and there a minnow and a tadpole swam among the weeds. They were inseparable friends.*
>
> *One morning the tadpole discovered that during the night he had grown two little legs.*
>
> *"Look," he said triumphantly. "Look, I am a frog!"*
>
> *"Nonsense," said the minnow. "How could you be a frog if only last night you were a little fish, just like me!"*
>
> *They argued and argued until finally the tadpole said, "Frogs are frogs and fish is fish and that's that!"*
>
> Leo Lionni, *Fish Is Fish*

With delightful illustrations, *Fish Is Fish* by Leo Lionni tells the story of these two inseparable friends—and their inevitable separation. The tadpole who develops into a frog departs the pond for a life on land. Meanwhile, the minnow grows and, of course, remains a fish.

Faithful amphibian friend that the frog is, he returns to the pond a few weeks later. The frog extols the wonders of land beyond the pond. As he shares descriptions of birds, cows, and people, the minnow imagines their extraordinary appearance: winged birds with legs, four-legged horned cows that "carry pink bags of milk," and two-legged people dressed in fine suits.[1] Colorful images form in the little fish's mind's eye.

He wistfully pictures a spotted fish with an udder (the pink bag of milk) and horns, a winged fish with two hind legs (the bird), and a walking pair of fish people, one sporting a black bowler hat and the other licking a lollipop. They all have fishy eyes, fishy heads, and fishy fins—plus limbs, the novelty that so impressed the frog. Imagining the world his frog friend has seen keeps the fish awake speculating about what his life would be like if he could leave the pond and "jump about."

Surprisingly, the minnow's fishy vision rings scientifically true. The bodies of all backboned creatures living on land resemble fish in significant ways—what paleontologist Neil Shubin refers to as "your inner fish."[2] Differences among backboned creatures are built upon the basic fish blueprint: mouth, head, two eyes, left and right sides, segmented spine, lots of ribs, and limbs in place of lateral fins. Gene mutation and fin adaptation propelled the descent of landlubber progeny from aquatic ancestry. The fishy ancestors of cows and people and birds did, in an almost literal sense, walk out of water onto land quite a long time ago, though not all at once of course. And the ancestors of these ancestors—beastly armored fish—sported modified pelvic fins as claspers for copulation.[3] Fins through time can do many things.

The legacy of limb fossils tells of fins turning into feet in tandem with the history of swim bladders becoming lungs. *Fish Is Fish* hints at this story on two levels: tadpole becoming frog and fish whimsically imagining land animals as modified fish. A frog needs lungs and legs; a tadpole, gills and a swimmer's tail. As the minnow watched day by day, his tadpole friend turned into a frog and left the pond. The challenges of the transition from aquatic to terrestrial habitat are striking: avoid drying up, overcome gravity, and evade sunburn.

On land their bodies must pump air in and out of protected chambers, propel their weight without a buoyant assist from water, and provide defense against the elements. If expecting to have any offspring, frogs must return to water to deposit (or fertilize) eggs. As tadpoles metamorphose, they metabolize their own tails and sprout limbs, hind legs first. They also prepare for a new diet (often moving from algae to bugs) and a new way of obtaining oxygen: breathing with lungs rather than gills. In the air they absorb some oxygen through the skin. It's a phenomenal change in the life of an individual creature. No wonder legs so entranced the minnow's froggy friend.

The analogous story of the minnow's ancient ancestors coming ashore unfolded over a much longer timespan. The story begins with one of the minnow's distant cousins.

The Lungfish at the End of the Zoo

Down by the Oregon Zoo's indoor Bamba du Jon (African swamp), crocodiles swim and thunder booms. Snug in its tank, the African lungfish's (*Protopterus annectens*) slimy, lithe body unfolds. The lungfish at the end of the zoo has abided 400 million years (well, not this one, of course—it represents the lungfish group), waiting for its story to be known. Like the minnow's tadpole friend, the lungfish knows both pond and land—and lots of mud in between.

The body is blunt at the head and ends arrow-like; a flexible cylinder followed by a triangle, with a somewhat flattened marble in front. Each of its whip-shaped appendages tapers to a point. Are they fins or limbs or something in between? It's scaly and can tolerate both the wet and the dry, the perfect portrait of a true survivor. And obviously it has lungs.

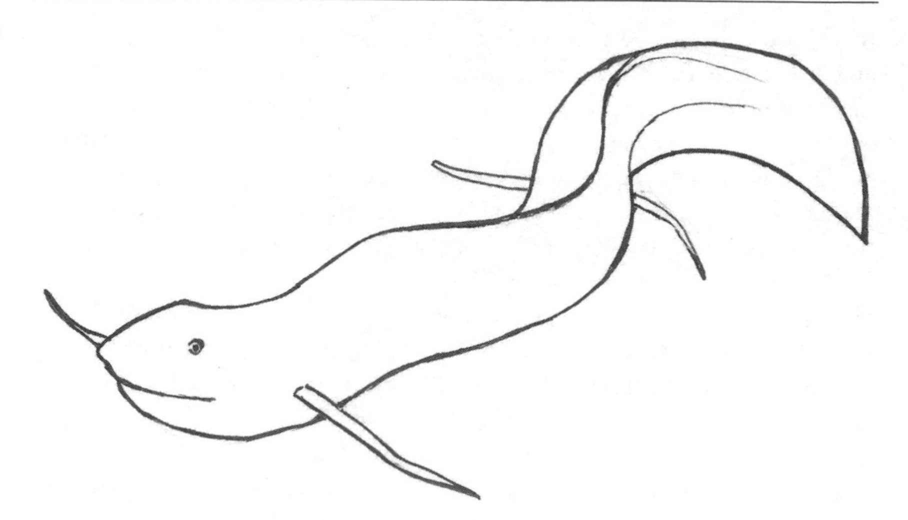

Figure 5.1 The West African lungfish (*Protopterus annectens*) at the end of the Oregon Zoo. Note the whip-like lobe-fins. Illustration by Jan Glenn.

The origins story of four-limbed creatures (e.g., frogs, cows, birds, and people) oddly resonates here. Limbs do remarkable things: hold bear claws, paddle seals, roost bats. They swim, gallop, grasp, climb, dig. They all begin socket-jointed to the body, then end in sharp claws, cloven hooves, and small fingernails. And they would appear to have started as lobe-fins on a long-extinct variation of (and cousin to) the lungfish. Among the tetrapods (all four-limbed back-boned creatures), there is always just one limb bone connecting to the body, whether at the hip or shoulder socket; then come multiple bones arranged in parallel style as the limb grows. There are always just four limbs among the backboned animals, whether arms, legs, or wings.

Limbs suggest natural groupings according to similarity of structure: those ending in an odd number of toes, and those ending in an even number of toes, for example. Assuming a long-ago tetrapod ("four-footed") ancestor had limbs that were more general in form than all of its descendants' brings the African lungfish to mind. A lobe-finned creature, similar to the lungfish, strode—or, more accurately, swam, scooted, crawled, paddled—earth's stage as the curtains parted for backboned life to walk on land.

Something Fishy

From egg to birth, embryo development among many creatures resembles the transformation of tadpole to frog. The changes mirror steps along evolution's long journey from fish to people. One reason for this similarity is the way genes control the development of limbs. To a surprising degree, torso-to-toe development resembles, at a genetic level, the body's head-to-tail design.[4]

Scrutiny of limb likenesses among fish and fish cousins, fossil and living, leads to counterintuitive claims of kinship. Anatomy persists from generation to generation, thanks to inheritance, but varies in some measure from individual to individual. Closer relatives inherit more similar anatomies than distant ones. Salmon and minnows are very closely related—and both are obviously fish. "Fish is fish" after all. Now the limb-driven punch line: cows and lungfish are more closely related to each other than either is to salmon or minnows.[5] The presence of shoulder and hip girdles with sockets and ball joints gives a measure of credence to this claim. But all fish, except for the limbless "jawless fish" (cyclostomes, such as the hagfish), sport shoulder and hip girdles. Those of the lobe-fins (lungfish included) appear a tad more tetrapodish. That's because they support the bony lobes attached to them.

Cows pictured in the mind of Leo Lionni's minnow had fins and fish bodies to boot. The minnow imagined cows not just as fish like himself but as fish with plump pink bags, four legs, and two horns. That image is close to the mark, given a substantially expansive concept of a fish. Cows are modified lobe-finned fish, something fishy yet without fins. Fish can be more than just fish.

Although excluding lobe-fins from the realm of fishdom may make sense to evolutionary biologists, in every sense of their shape, habitat, and behavior, lobe-fins are fish. Minnows are indeed fish of a different sort. Minnows, trout, guppies, tuna, and the like are stereotypical fish to most people. In the fins of these fish thin skin stretches over rays of bony or horny spines. They have no fleshly lobes.

Ray-fins and lobe-fins do share a common ancestor from a very ancient era when primitive armored fish—placoderms—dominated the seas. Both groups have jaws, so that means they share—along with cartilaginous fish (sharks and rays)—an armored, jawed ancestor. Certainly, these armored Methuselah creatures counted as some kind of fish. Fossils of the giant Devonian predator, *Dunkleosteus*, attest to their reign.

Admitting that salmon and minnows are one kind of fish among many is easier than accepting lungfish and cows as another kind of fish. Swallowing cows as "fish"? That would mean facing up to the culinarily awkward idea of making fish sticks from hamburger meat. For the evolutionist, however, cow-fish kinship occupies a branch on the tree of life, presenting a conundrum only from the perspective of meal planning.

Banishing the category "fish" from the business of naming altogether seems extreme and counterintuitive. Already the term "whale fish" has succumbed to disfavor. "Sea stars" instead of starfish works fine, but naming all shellfish "sea shells" causes confusion. Everyone knows a fish when they see one; "fish" is in the eye of the beholder. "Behold the cow, a mighty fish!" won't play in Peoria.

The human mind is indeed quite capable of deciding what makes a fish a fish, a frog a frog, and a cow a cow from the point of view of human purpose: eating, for example. "Fishiness" is a mental invention useful to people's navigation of the natural world in search of food. Dairy farmers do not milk fish, and cows start out as live-birthed calves. There are obvious, substantial, and important differences. No one sings "You get a line and I'll get a pole, honey; you get a line and I'll get a pole, babe; you get a line and I'll get a pole, we'll go fishin' in the . . . *cow pasture*." The lobe-fin inspired lyric fails to rhyme or make sense.

Nevertheless, people and cows share an inner fish inherited from a lobe-finned ancestor of long ago. And the lobe-fin, obviously a fish inwardly and outwardly, holds in common with minnows an even more ancient ancestral fish. Limbs do not lie; the language of DNA attests to deep truths of fish kinship.[6]

Today's human twigs on the tree of life stem from a branch living in Africa about 140,000 years ago.[7] Much longer ago, on the order of 570 million years, existed the distant ancestor common not only to humanity but also to starfish (sea stars). It's a fish quite a way back in time. All the lungfish, the other lobe-finned fish, the ray-finned fish, the four-legged land animals, the birds in the air, and the whales in the sea—and much more—spun off en route to the present.

According to Leo Lionni's minnow, "fish is fish." Yes, but there are a great many varieties. Very, very many varieties, modified during the long march of descent into forms most wonderful and beautiful to behold. The idea of "fish" serves many practical purposes, such as distinguishing hamburger from fish fillets and avoiding confusing the responsibilities of fishing with those of herding cattle. "Fish" functions best when preceded by another category: shellfish, starfish, shark fish, ray-finned fish, crustaceous fish, squid fish, and whale fish. It implies creatures of the sea. These useful categories, however, provide little insight into ancestral affinities. Perhaps Neil Shubin should have referred to our "inner lobe-fin," but inner fish will do if we keep in mind that fish is not necessarily fish.

Ball-and-Socket Girdles

The antiquity of the lungfish implies an equal or greater antiquity of nostrils and lungs. The strange fins of the African lungfish (and the South American lungfish, too), attached limb-like to the body, look more like whips than fins. That's not true for the Australian lungfish and the majority of fossil lungfish species.[8] When the little minnow imagined forms of life on land, he imposed his fish body on each one and skipped over crawling. A fish that crawls might not have recognizable legs and feet, but that's a place to start. It should have parts that grow and arrange themselves in a limb-like pattern.

Having a lung, not to mention nostrils, would seem a bit unusual for a modern fish, unless that fish is a lungfish. Might a lungfish walk into a zoo? Probably not;

its fleshy fins house strings of rigid elements anchored to hip and shoulder girdles, but there are no ankle or foot structures. It might scoot along, but it's no tetrapod. Dry up a river and the lungfish will excavate a burrow in the mud and wait for the water's return, nicely nestled and breathing peacefully.

The lungfish has paired, fleshy fins, both pectoral and pelvic (shoulder and hip), and the rudiments of ball-and-socket joints found where legs and arms meet torsos, yet with the pattern among the known tetrapods reversed. The ball on the top of a frog's leg bone, a human's leg bone, or the lobe-fin of a coelacanth (the celebrated living lobe-finned fish) fits neatly into the socket at the shoulder or hip. The first bone in the body of the lungfish's fins differs. It contains the socket. The rounded form of the torso girdle anchors the socket on the fin bone.[9] Two similar designs—a genetic coin flip apart—provide for movement with strength among lobe-fins and their progeny.

One lineage settled on one pattern, its cousin on the other. With the ball-and-socket plan, fish fins acquired good leverage and great range of motion. Incipient crawling had found its footing. For over 400 million years, variations have endured on the double-lung, fleshy-finned, predator-toothed, flexibly backed lungfish body plan and in its sister creatures' ball-and-socket mirror image.

Ball-and-socket limbs and girdles and lung-breathing bodies are traits well suited to land lubbery. Yet clearly these features worked perfectly fine right where they started, in the life aquatic. The lungfish still earns a living according to the ancestral approach. And it breathes.

Lungs Are Lungs

The living lungfish lacks a pink bag of milk, wears no clothes, and cannot fly—as do the land fish in the imagination of Leo Lionni's minnow. On the one hand, it cannot stroll about its terrarium at the zoo nor rear up on its hind fins in an aggressive posture. On the other hand, this air-breathing fish is a piece of the puzzle a glimpse of finned life poised to take on the land, from a time when fish with gills and only gills were perhaps the exception, not the rule.

Today's lungfish have multi-pocketed lungs with connections to external nostrils on top of their heads, the better to breathe when submerged in sunbaked mud. The ancestor common to our inner fish lineage and all lungfish probably had a flattened head and paired nostrils too. Ancient lungfish, along with their lobe-finned fish cousins, thrived in deep time. From shoulder to finger or hip to toe, bones of the lobe-fins spun off a host of forms, yielding the tetrapod life (amphibian, reptilian, avian, mammalian) of today. Look at the feet of a crocodile or even a salamander, then at your hands. They are strikingly similar; the bones of the forearm and upper arm even more so. Remove the fleshy pads from an elephant's foot and the same pattern appears. It's really quite eerie—that inner lobe-finned fish in us all.

The frog lost his tail and his gills while growing lungs in preparation for life on land. Gills and lungs are different in very many respects—as different as fins and feet. Gills come and go as embryos become adults among tetrapods, but gills do not transform into lungs. Lungs are lungs.

Limbs (and fins) work nicely when powered by well-oxygenated muscles. To oxygenate those muscles, lungs absorb oxygen directly from the air. Gills absorb oxygen that has dissolved in water. Whether from water or air, the essence of breathing is gas exchange: oxygen absorbed and carbon dioxide released. The structure of both lungs and gills multiplies surface area in order to maximize the absorption of oxygen. Gills branch and branch; lungs fold and fold. Both patterns create more surface area. But why do spongy lungs work so well on land and feathery gills so poorly? First, gills dry out too easily; lungs enclose the moist, oxygen-absorbing surfaces. Second, air-breathing takes advantage of the fact that the concentration of oxygen in the air is much greater than in the water. Air-breathing with lungs makes more oxygen available for the blood to deliver to the muscles. Therefore, more fuel may be burned faster. Lungs energize the blood.

Because they are enclosed and easily sealed, lungs are good for something else: floating. In water, holding a breath of air adds buoyancy; completely exhaling makes sinking inevitable. Breathing air works only at the surface, and inhaling air keeps you there.

While the frog in Leo Lionni's story was busy sprouting legs, its minnow friend enjoyed roaming up and down throughout the pond. People rise and sink by holding their breath or blowing all of the air out of their lungs. Perhaps ancient lunged lobe-finned fish did the same thing. Lionni's minnow compressed and inflated its swim bladder, aided by a gulp of air.

Many kinds of fishes willfully change buoyancy to float or sink. They control their depth by inflating and deflating (or squeezing and expanding) an internal swim bladder, a lung-like organ. "Hold your bladder" means something entirely different to a fish than to a person.

Lungs—enclosable chambers kept moist for breathing dry air—have surprising similarities with swim bladders and vice versa. Swim bladders—enclosable chambers that swell and shrink to change buoyancy—have surprising similarities with lungs. Lungs themselves can function to aid floating. Nevertheless, the lung-to-bladder switch in function is puzzling indeed. Both begin to form as pockets in the gut during the development of the embryo (or tadpole) but offer advantages in different circumstances.

Which came first, the swim bladder or the lung? Their similar development troubles the problem of which evolved first, the swim bladder or the lung. Perhaps the swim bladders of fish are modified lungs rather than, as is oft supposed, the other way around.[10] Primordial bony fish breathed air with a lung-like organ, and therefore the tadpole's minnow friend likely came from ancestral stock that could grab a bite of air. Many air-breathing fish also have the capacity to extract oxygen from water directly. Therefore they don't drown without that bite of air. The gulp of oxygen is a valued extra. The puzzle, therefore, is figuring out the benefit to a fish of a bite of air for life underwater.

A Problem with Fish Hearts

What might have been the circumstances favoring fish lungs? The benefit of a gulp of air for life underwater makes sense in the context of how a fish heart works. Lacking lungs, and limited to breathing water by means of gills, a fish has reduced access to oxygen. Oxygen depletion strains the heart and limits muscle power. Fish hearts pump blood first to the muscles, then on to the heart *after* the blood has circulated through the muscles. That means that the blood reaching the heart may be depleted in oxygen. An extra gulp of air helps to oxygenate the blood and hence energize the fish's beating heart, as Carl Zimmer explains in *At the Water's Edge*. That solves one puzzle: how lungs benefit fish. They aerate the heart.

Bursts of energy expended by predatory fish tax the heart. The hunter gulps for air. Yet only a very small percentage of fish living today breathe air directly, and even in oxygen-scarce waters most fish depend solely on water-breathing through gills.[11] Very many fish hunt prey. So the first solution (aerating the heart) sets up a new puzzle. Why are air-breathing fish such as the lungfish such rare curiosities in modern waters?

Zimmer suggests that thoughts about predator-prey behavior help to resolve the dilemma. Fish wish to avoid being a meal while, in many species, wishing to make meals of other fish. If predators lurk near the surface, deep water may offer safety. There's no gulping air down there, and the predator may tire of the hunt. For potential prey, the alarm call of "Dive! Dive!" sounds when a predator lurks above. Go deep, and go quickly. Swim bladders enable potential prey fish to submerge and rise at will with little expenditure of energy. They use the energy spared to flex muscles and flee. The frightened prey fish shrinks its bladder, sinks fast, and swims like the dickens. To catch them, predator fish must have a swim bladder too.

Gulping air kept the hearts of ancient predatory fish beating strong and their muscles well oxygenated, but only while they prowled near the surface. Gulping air to power surface prowling made these predators quite shallow creatures. Diving deep became a good way for the minnow's ancestors to escape from predators lurking near the surface. Given that flying predators swoop down from the air to grab prey fish from surface waters as well, quick-diving made added sense.

When the primitive predators themselves dove deeply—aided by neither swim bladder efficiency nor an air-breathing assist—they risked cardiac arrest. For the most part they became extinct, and the swim bladder's energy savings formula triumphed.[12] Sink and swim rules predator and prey behavior to this day.

Throughout the second half of the twentieth century, most biologists accepted the argument that the first jawless armored fish (the placoderms) lived in freshwater and depended upon rudimentary lungs to survive. Reinterpretation of placoderm fossil remains, however, has recently challenged that view. Daniel Goujet, a distinguished paleontologist at the Muséum National d'Histoire Naturelle in Paris, has discredited this idea. First, he noted that presumed lung-like structures were more likely distorted remains of viscera once part of the digestive tract. Second, he demonstrated that the environment inhabited by *Bothriolepis*, the widespread, supposedly lunged fossil placoderm known from the Middle to Late Devonian, was not a shallow freshwater creature but rather a marginal marine one. He now refers to the notion of lungs in placoderms as a "persistent paleobiological myth."[13]

Thus the origin of lungs among ancient groups of fish remains problematical. Following upon the diversification of lobe-fins, legs and lungs tended to march together through evolutionary time, modifications in one enhancing those in the other, opening new nooks and crannies for fishy things in swampy, log-choked, flood-prone wetlands. Some fish did (and still do) sport small bones within lobes that attached their fins to their torsos (the lobe-finned body plan). Some fish did (and still do) breathe air with lungs. Lineages destined for life on land, or life lived between land and water, clearly kept the advantageous lung and developed the limb pattern to the max.

What has become of the swim bladder? Sturgeon and teleost fish have them, as do coelacanths; gar and bowfin even breathe with them. Teleosts include stereotypical fish—perch, bass, and minnows, for example. Those who group fish according to similarities due to inheritance nest teleosts with other ray-finned fish (sturgeon, gar, bowfin) into a diverse array of bony fish with and without spines in their tails. All of the many thousands of species of fish whose spine ends before the tail fin, thus presenting a fanlike appearance, are teleosts. (The tail bones continue along the top of the tail fin in the sturgeon.) Most modern teleosts are bony, scaly, swim-bladdered, and lungless—as respectable "true" fish should be. Yet some of the earliest teleosts to evolve call upon their swim bladders for respiration. Nothing prohibits the evolution of dual function. Teleost is teleost.

When people mimic fish, they tend to move their mouths in the manner of the protruding jaws of a teleost. Anchored to the skull, both upper and lower teleost jaws are movable. Originally, moving jaws may have helped fish pump water through their gills or air into their lungs.

Splitting the lobe-fins from the ray-fins marks a dramatic distinction, more basic than the presence of lungs. Both lineages have calcified bones (although the skulls of modern lungfish are cartilaginous), unlike sharks and rays, more distant cousins still.[14] Sharks lack lungs and depend on cartilaginous "bones" for structure. Nor have they swim bladders: sharks' oil-rich livers help them stay buoyant. No ribs, either. Given the intermediate forms connecting lobe-fins to tetrapods and the incredible diversity of body structure among fishes (jawed and jawless, bony-armored and scaled), there seems to be a slim chance of deciding unequivocally whether fish is fish or not, as Lionni's minnow so confidently asserted.

Lobe-finning forms branched early on, becoming lungfish on one branch and tetrapods on the other. Descendants of lobe-fins experimented with nostrils on top of flattened heads, the better to breathe with; neck bones separating head from shoulders, the better to snatch prey; and digitized limbs, the better to crawl about in swampy environs. Amphibian progeny took to the land but retained watery hatcheries and nurseries. Reptilian lines conquered aridity completely with skin and scale, eggshell or live birth. Birds and mammals happened.

Tiktaalik and the Dawn of Tetrapods

Ordered in time, fossils illustrate the path from lobe-fin to bony limb quite convincingly. Where shales outcrop in the northern lands of Greenland, Ellesmere Island (Canada), Quebec, and Pennsylvania, paleontologists have searched with striking success for pages of the fossil record written in tetrapod script.[15] The trail of fossils entombed in Canada's shales and their corresponding dates fit an expected sequence from the undifferentiated to the specialized. *Eusthenopteron*, a lobe-finned elder, swam in deep-water Devonian seas during the aptly named "Age of Fishes."

As explained by Carl Zimmer, the sequence of bones in the lobe-fins of *Eusthenopteron* corresponded to the limb pattern of a single bone, then double bones, then many small bones—one, two, a few—typical of tetrapod creatures.[16] This pattern has persisted in the tetrapod forelimb for tens of millions of years. An upper-arm humerus attached to the shoulder. The lower arm's radius and ulna extended from the humerus. Wrists and digits completed the "many" component of the pattern: torso joint, single bone, double bones, lots of bones.

In the tetrapod hind limb, the single thighbone or femur was followed by the lower leg tibia and fibula. Ankle bones extended from the side-by-side tibia and fibula, ending in toes. The tibia and fibula attached to the femur. Sung at many a campfire, the lyric "Thighbone connected to the hip bone" proves that knowledge of the tetrapod limb pattern is indeed widespread.

Fossils illuminate both the bookends (lobe-fins and tetrapods) and the transitional fin-to-limb titles shelved between them in the shales. Tetrapod creatures, exemplified by *Ichthyostega*, retained the single, double, many-bone pattern pioneered by lobe-finned fish such as *Eusthenopteron*. Some 365 million years ago, eyes atop its head alert for prey and ready to ambush, *Ichthyostega* sprawled on true feet.

 Sandwiched in time between *Eusthenopteron's* lobe-fin locomotion and *Ichthyostega's* pouncing stance reside the fossils of the truly transitional *Tiktaalik* and the early tetrapod *Acanthostega*, dated 10 million years apart. These intermediates grew forelimbs terminating in zero and eight toes, respectively. *Tiktaalik's* foot-not-yet-a-foot and *Acanthostega's* excessive toe count suited them for sloppy walking at best.

On April 6, 2006, the journal *Nature* introduced the world to a fossil superstar: skeletal pieces of *Tiktaalik roseae.*[17] The name, proposed by Inuit elders, simply means "large freshwater fish."[18] Found amidst rocks 375 million years old on Canada's Ellesmere Island, "this fish could do push-ups."[19] Its discoverer, Neil Shubin, proved nature to be whimsy's equal and maybe its superior. With its "pectoral appendage" (fin in the shoulder position), which was "transitional between a fin and a limb," *Tiktaalik* "was capable of a range of postures, including . . . a stance in which the shoulder and elbow were flexed."[20] Rarely do people think of fish flexing their shoulders and elbows (a frog, maybe), though that is certainly something Leo Lionni's little fish might have imagined.

In January 2014 Shubin debuted astonishing drawings from his discovery of *Tiktaalik's* fossil pelvic girdle. Its hip sockets anchored lobe-finned paddles capable of a wide range of motion and perhaps some footless walking in the manner of the living African lungfish.[21] *Tiktaalik* had ball-and-socket joints fore and aft, but no toes.

On the other, more recent side of *Tiktaalik's* date of 375 million years ago, paleontologists discovered tetrapods with obvious toes, yet still with rather long tails. At 365 million years ago, between the dawning of the lobe-finned *Eusthenopteron* and the ambulations of the leggy *Ichthyostega*, lived *Acanthostega*. It clambered clumsily about, close on the finny heels of *Tiktaalik* of push-up fame. *Acanthostega* shared with its pre- and post- cousins a single humerus at the shoulder, followed by the double bones of radius and ulna. Instead of being of nearly equal length, as is the norm for tetrapod walkers (and human forearms), they closely resembled the lobe-finned pattern: short ulna, long radius. *Acanthostega* took the next step of the sequence ("then many" in the pattern) rather seriously. Its bunch of wrist bones terminated in eight toes. An unbendable elbow limited its forelimb. Imagine tracking a critter leaving eight toe prints on each front foot track.

In the figure, *Eusthenopteron*, a lobe-finned fish from 380 million years ago, lurks in the water. *Tiktaalik*, an intermediate between lobe-fins and tetrapods dated at 375 million years ago, perches on the shore. According to Shubin, *Tiktaalik*, the star of this show, could bend its neck, a trait that continued to evolve in subsequent tetrapods. The diagram omits *Ichthyostega*, a semiaquatic early tetrapod from 365 million years ago, and next in the fossil timeline. A generalized reptile-like amphibian from about 320 million years ago walks near a tropical plant. The obvious dog and human complete the sequence. The bones of the forelimb that correspond from one creature to the next are highlighted.

Figure 5.2 Similar bones in similar places in the vertebrate forelimb. One, then two, then many from upper arm to forearm to wrist to digits. Illustration by Jan Glenn.

Genetic Blueprint from a Living Fossil

A veritable zoo of lobe-fins, proto-tetrapods, and archaic tetrapod fossils peppers the rocks from Devonian to Carboniferous time—lung-laden all. Two lineages of lobe-fins evolved species that exist today. They are relics in form from the dawn of limbs and lungs, although they have likely changed on evolution's timescale in ways that fossil skeletal remains cannot preserve. Most notorious is the coelacanth, a lobe-fin thought to have been extinct for 70 million years until it introduced itself to modernity when one was caught off the coast of South Africa in 1938. Actually, coelacanths had probably been caught before, but without being brought to the attention of the scientific community.

Coelacanths are more distant tetrapod relatives than are lungfish. They give living flesh to the very ancient lobe-finned fossil form. The African coelacanth (*Latimeria chalumnae*) has maintained the primitive lobe-finned body pattern for 300 million years. A second species of coelacanth was discovered in 1997 in Indonesian waters (*Latimeria menadoensis*).[22]

In Devonian time, flat-snouted, spread-nostriled *Tiktaalik* crawled into the picture, and backboned terrestrial life followed in its tracks, facing challenges of locomotion, desiccation, and open-air metabolism. Limbs, eggs, and kidneys adapted. Though not as closely related to living tetrapods as the lungfish, the coelacanth holds clues about the transition from lobe-fin to land animal.

At least that's what recent genetic data confirm. As summarized in a 2013 report published in *Nature* by an international research team led by Chris Amemiya, "the modern coelacanth looks remarkably similar to many of its ancient relatives, and its evolutionary proximity to our own fish ancestors provides a glimpse of the fish that first walked on land." Knowledge of the coelacanth genome to has the potential to decode the genetic story of tetrapod evolution and the transition by vertebrates to life on land. In brief, the coelacanth genome is "a blueprint for understanding tetrapod evolution."[23]

Changes in genes controlling nitrogen excretion, immunity response, and limb and organ development may stand out in the comparisons between coelacanth and living tetrapod genomes. Comparisons may also identify genes governing the origin of tissues that protect the developing tetrapod embryo from the stresses of terrestrial birth. The Amemiya team successfully sequenced the complete coelacanth genome as well as samples from the genome of the West African lungfish (*Protopterus annectens*) obtained from its brain, liver, kidney, gut, and gonads. By comparing the relative similarity of coelacanth and lungfish genes to genes sampled from twenty-one contemporary jawed vertebrates (e.g., platypus, armadillo, elephant, dog, mouse, turkey, puffer fish, spotted catshark, human, and more), the team concluded that "the lungfish, and not the coelacanth, is the closest living relative of tetrapods."[24]

Ode to the African Lungfish

Lungfish, as already suggested, hold a special place among creatures of demonstrated antiquity. Humans share elements of the blueprint for its body. In terms of kinship, lungfish claim sisterhood status with other lobe-fins, the coelacanth being a bit more distant lob-finned cousin. Forward in time, the non-coelacanth, non-lungfish lobe-finned branch begets a host of tetrapods. Early, the amphibians branch apart from all the other backboned creatures who walk, slither, and run.

People walk on hind limbs ball-and-socketed to their frames. In ancient Devonian waters, ancestors to us and our lungfish cousins roamed near the shores, their pectoral and pelvic lobe-fins joined to torsos with bones destined to empower walking. The ends of our lobe-finned appendages are so specialized we've renamed them hands and feet. Attaching our very human hands to our fish-descended bodies are bones with the names humerus (upper arm), radius and ulna (forearm), carpals (wrist), metacarpals (palm), and phalanges (fingers). These bones have changed through hundreds of millions of years, yet the one, then

Figure 5.3 A lungfish walks into the zoo while its cousin purchases a ticket. Illustration by Jan Glenn.

two, then many pattern held constant. Humerus stays humerus, and the ball-and-socket plan endures.

Stare at the African lungfish at the end of the zoo while the lungfish stares back—two lobe-fins with a common genealogy eyeing each other. Are we the fish this fish wishes to be? Do we see in this fish the fish that we were? We're different now, yet still alike. A book title from childhood memory crosses the mind: *Fish Is Fish*. A lungfish walked into the zoo was it you?

An Ode to the African Lungfish

A body lies snug in its mud-crusted lair,
As ancient eyes gaze in vacant stare.
Here lies a slithery thing, slimy head to tail:
A non-descript tube, like an uncoiled snail.
Fleshy fin-limbs stretch out from its slippery side,
Stubby whips that assist when it glides.
Does it know its affinity to those that breathe air,
To the tetrapod creatures who inhale through their nares?
One tiny bone, encased in a lobe,
Joins fin to body, then more segments unfold.
When it moves, does it crawl?

Does it ambush at all?
A tetrapod cousin with scimitar teeth,
Shares the bony limb structure bequeathed
To the backboned contingent all over the earth—
Whose lobe-finned origins preceded their birth.

6

OUT ON A LIMB

Sketching Bone by Bone from Joint to Joint at the Zoo

And she sang the low, crooning seal song that all the mother seals sing to their babies:
You mustn't swim till you're six weeks old,
Or your head will be sunk by your heels;
And summer gales and Killer Whales
Are bad for baby seals.
Are bad for baby seals, dear rat,
As bad as bad can be;
But splash and grow strong,
And you can't be wrong.
Child of the Open Sea!

Rudyard Kipling, "The White Seal"

Readers of Kipling's lovely poem may find themselves taken aback by the idea of seal heels. Most often a seal's forelimbs are described as "flippers." Its hind limbs—in effect, a second pair of flippers looking much like a tail—do have what might be called heels. Swimming with flippers is something baby seals need to learn lest they fall victim to an ocean predator, the killer whale.

An arms and legs race has unfolded between predator and prey ever since the lobe-fins wiggled ashore to become tetrapods. Some returned to the seas: the flippered and the fluked, the white seal and the killer whale. From pinniped to pachyderm, limbs differ dramatically in order to accomplish crucial functions, and zoos are well suited to the study of this diversity. Creatures held captive in zoos or animated with exaggeration in cartoons can get close looks, affording an investigator the opportunity to figure out whether, for example, elephants have knees or seals have heels—and maybe even hips. The trick is to draw them.[1]

Knowledge of comparative anatomy, especially limb structure, served cartoon animator Chuck Jones very well. Jones, creator of Bugs Bunny, Daffy Duck, Wile E. Coyote, and many other familiar Looney Tunes characters, perfected both form and movement in his animations. Jones clearly anthropomorphized the faces of his creations, but he also maintained accurate limb proportions and kept limb joints in their proper configurations. A powerful hop in an unanticipated direction is a good way to escape from a predator. Bugs Bunny, to the consternation of Elmer Fudd, embellishes his escapes with numerous tricks—but he still runs on proper bunny feet. The speedy Roadrunner, abetted by similar trickster stunts, forever seems to frustrate Wile E. Coyote, but likewise does so on bird limbs that are scientifically accurate.

In his autobiography, *Chuck Amuck*, Jones recounted the story of animating Rudyard Kipling's tale "The White Seal" as a Disney production. Let's forgive for a moment the animator's failure to distinguish between seal and sea lion. What the Disney people expected was animation that captured pinniped swimming (seals, sea lions, and walruses) in a convincing manner.

Sea lions do have several traits that distinguish them from seals: small external ears, for example. Sea lion swimming parallels seal swimming, however, because both animals utilize similarly structured flippered or "finned" feet, modified appendages descended from their land-living forebears.

Pinniped Grandchildren

Jones understood the relationship between structure and function. He realized that "all animals move the way they must move, because their unique anatomy develops as necessary in each unique environment, and the sea lion is no exception." To improve his animation, Jones went to the San Diego Zoo, where he studiously watched sea lions:

> I could not conduct a course in comparative anatomy with my animators. I was not educationally equipped to do so; therefore, I followed the most logical substitute—comparing my own anatomy to that of the sea lion. This is not as difficult as it might sound, since all vertebrates have more structural matters in common than differences. Our bones and muscles all bear pretty much the same names and are readily identifiable; the great differences are primarily in length and weight of the bones and the musculature, and, of course, in the skull structure.[2]

He observed that onshore, sea lions moved with a "complacent wobble" similar to the way "movie moguls actually walked." Once in the water, everything changed: "The sea lion becomes a sinuous master of the aquatic arabesque, a series of graceful notes swirling through the water with confident beauty."[3] Jones concluded that the sea lion's flippers were hands and feet, not fins, because they had obvious toenails and fingernails. He guessed that the upper arm bone (humerus) remained within the body. Only the lower arm (radius and ulna plus wrist and hand) from the elbow joint outward emerged. How did Jones test this hypothesis without conducting a sea lion dissection?

Back at home and away from the zoo, he recruited his two grandsons, ages eleven and thirteen. Next he tied their arms to their bodies down to the elbow and tied their legs together from hip to ankle. He helped them don swim fins on both their hands and their feet. Apparently, boys with arms pinned to their bodies and legs bound together work perfectly well as analogues to pinniped mammals.

Had the seals been present, they would have looked on with amazement, bewildered at the site of boys imitating seals, but relieved that such boys might well serve as stunt doubles. Trained seals are willing to go only so far in helping with the filming of a Disney production before a stunt double has to be called in.

Jones's animator colleagues assembled as an audience, and into the pool tumbled the two highly animated grandsons. "Within minutes they were swimming the only way they *could* swim—awkwardly, but exactly as a sea lion swims." They dove in sweeping arcs, though less gracefully than true sea lions, and surfaced quickly to breathe—in a much greater hurry. "They were as close as a human being could be to a sea lion, and the awkwardness of their movement could easily be corrected by the animator."[4]

Yes, this was a looney idea. Do not try this experiment at home with your own children or grandchildren, or other people's children for that matter. In fact, it would be a bad idea to try this experiment on yourself, especially with no lifeguard on duty. The good news: the boys survived just fine, although they spent the rest of the afternoon bouncing beach balls on their noses.

Shoeing the Ankle

In *Chuck Amuck*, Jones's sketches of Elmer, Daffy, Bugs, and Porky Pig appear on the same page as his sketches of horse, human, dog, cat, and cow limbs. The caption states, "Same bones—different lengths only."[5] Jones's drawings superimpose the body forms of the cartoon characters on lines corresponding to curves, angles, and orientation of the backbone and limbs positioned for motion. Change the angle, change the length, change the character, and change the action.

In another illustration, Daffy, Bugs, Elmer, the Tasmanian Devil, and Yosemite Sam are in a police lineup. The first frame uses horizontal lines and vertical axes to establish proportions. The next frames add basic shapes, then body contours. In the last frame, Jones superimposed photos of his animator colleagues.[6] They probably chuckled when they saw the result.

Humorous, yes, but more important, Chuck Jones's cartoons teach how to invent new, strange, and silly creatures by varying the proportions, angles, and basic shapes used to draw real animals quickly. His techniques keep these creations believable as skunks, ostriches, and roosters.

With just modest practice, even novice artists may succeed in rendering realistic drawings of zoo creatures. A good training exercise in this regard is to play Nike designer to the tetrapod kingdom. The Chuck Jones challenge is to design a pair of sneakers that properly fit a horse, a kangaroo, or a sloth.[7] Recognizing how and where the joints of the foot and leg bend are essential to this task. Think length, angle, action. The trick, Jones explains, is to know which parts of the foot, ankle, and shin belong inside the sneaker and which parts do not.

On a human, basketball shoes cover the ankle. A Nike designer will want kangaroo, horse, and sloth sneakers to do the same. If a kangaroo were to play basketball while wearing the proper footgear, the shoe, as Jones realized, must tie just above the ankle. Where's the ankle (tarsal bones) on a kangaroo? It's just above the back part of the foot that touches the ground (the heel or calcaneus) when the kangaroo is at rest. That's the same manner in which the human foot touches the ground when at rest. A standard Nike basketball shoe, size one hundred, would be a good approximation of a sneaker fit for a kangaroo.

Socks look sharp when pulled up to the knee, placing the venerable swoosh in clear view. So, on which part of a kangaroo leg do you draw the sock? The kangaroo's knee joint lines up with its belly. It's going to need an extra-long pair of tube socks.

The challenge of fitting a sneaker to a three-toed sloth or a draft horse is somewhat greater. In the sloth, the joints are more subtly hidden by the blobby body, but the comparative strategy suggests, joint by joint, where to look. The sloth sneaker will most definitely require strongly reinforced toes, given its long, curling claws.

Horses, of course, have a joint that corresponds to the human knee. It joins the base of the thigh bone (femur) of its hind leg to the top of its tibia. The patella, or kneebone, caps this joint, known in equine circles as the "stifle." The ball end of the femur fits into the socket of the hip. For the most part, the femur section is at belly height and above, the knee (stifle) thus placed well above the ground. The leg bones swing at the hip, bend at the knee, and angle again at the ankle joint. The horse's ankle (its tarsals and calcaneus or "hock joint") sits rather high as well. When a horse is shod with a human shoe, the sole runs from the toe to just beneath the ankle, giving the horse sneaker an elevated sole. The horse actually perches on its toe and the single toenail at the end of each foot thickly expressed as a hoof. Maybe that's how the *Perch*eron got its name. (Actually, Perche is the name of a place in France where the Percheron originated.)[8]

Humans play basketball quite well with just one pair of shoes and naked hands. Because they have legs and not arms, horses will need two pairs. With its upper limbs free, a kangaroo might do okay on the court. Horses, however, probably cannot dribble or shoot well with their forelimbs. Still, you might wish to design front-foot sneakers for them and leave them to set picks—an awesome equine defense! Designing with analogy to humans in mind is more difficult because people do not wear shoes on their hands (except, perhaps, in silly skits at summer camp, making, quite literally, asses of themselves). To arrange a fit similar to the one for the hind limb, the forelimb shoe must enclose the knuckles (or "fetlock" on the horse) and reach to just above the horse's wrist (or "knee"), partway up the front leg.

Dog knuckles, wrists, ankles, elbows, and knees fall in between those of kangaroos and horses from the perspective of sneaker-fitting. If you are interested in turning a dog into a sea lion with swim fins instead of sneakers, it is best to start with a dachshund, urges Jones.[9] They might do better at water polo than basketball, given their height. At the very least, a comically swimming, finned dachshund will remind us of our lobe-finned ancestors.

At the Oregon Zoo, Eddie the Otter is the star basketball player.[10] He dunks with abandon. Of course, Eddie's brand of basketball is aquatic and hence his sneakers are just for show. They would get in the way of his webbed feet, needed to propel him through the water and up to the rim, dribbling water from his whiskered chin as he dunks the ball. Zookeepers have prescribed hoop-shooting as therapy for Eddie's arthritis. (When Disney decides to animate "Eddie's Hoop Dreams," the animators will be able to watch the otter on YouTube.) I imagine that Chuck Jones would have had more difficulty using his grandsons to model an otter's swimming than that of a sea lion. Otters propel themselves with large tails, which boys lack.

Elephant Knees and Giraffe Elbows

Imagine touring a zoo guided by Chuck Jones and his animator team. Enter with an artist's sketchbook in hand, asking, "How do limbs differ from creature to creature?" The task is to sketch the limbs of a variety of creatures from their shoulders and hips to the tips of their digits on the fore and hind limbs. A few exhibits may have models, sculptures, or skeletons that will help you do so. The previous chapter introduced the strangest "limbs" of all, the appendages of the lungfish in the African swamp exhibit. Crocodile limbs are quite interesting as well. (Do keep your distance unless seeking to turn your nose into a trunk while examining them—the story found in the next chapter). Crocs often remain still or move slowly, making the digits readily apparent and the angles at the joints easy to observe.

Sketching animal limbs from torso to toe tip engages zoo visitors in comparative anatomy and focuses attention on the positions of ankles, elbows, knees, and wrists. One amateur artist begins a sketch of an impala and immediately discovers "Its knees bend backward!" Another adds, "Looks to me like the stork's knee bends backward too."

Nearby, another zoo visitor looks at an elephant and wonders "Do elephants have knees?" "Yes," a friend responds. "They're right there by the belly and easy to see when the elephant walks." Anthropocentric perception prompts zoo visitors to observe that the hind legs of many creatures appear to "bend the wrong way."

At first glance, the supposed knee joint of a large quadruped mammal (especially the ungulates) or a bipedal bird may appear to bend backward. That's because its toe-perching stance and stretched-out lower bones lift the ankle joint (taken for a backward-bending knee) and place the knee high (the joint between the femur and the radius and ulna) and close to the torso. Humans grow accustomed to feet with heels on the ground. For animals that tend to stand up on their toes, the heel takes an elevated position similar to that of the human knee. It sticks out to the back like a knee that bends backward.

The long forelimb of the giraffe prompts a question analogous to the elephant knee conundrum: "Do giraffes have elbows?" In both cases the answer is yes, but convincing oneself that this is so requires tracing the bones from joint to joint, noting the angles as they flex, and placing segments of animal limbs in a sequence that corresponds with those of a human. Doing this makes good use of human-centric perception. Exhibit by exhibit, sketchers may proceed on the prowl to detect wrists, ankles, knees, elbows, heels, and toes.

The elephant's toe-standing stance may prove difficult to discern and counting its toes somewhat confusing as well because of the fatty pad in the back of the foot that provides support. Because zoos often lack exhibits of skeletons, try to check out an elephant or mammoth skeleton on your next visit to a natural history museum. A reconstructed skeleton will lack the fatty pad, of course, bringing the structure of the foot into clear view and making apparent the tippy-toe stance.

How strange to think of a kangaroo sitting back on its ankles while the bulky elephant tiptoes about.

Each elephant foot also has in the back a false sixth toe within the fatty pad. The sixth toe is a bony structure whose existence has puzzled scientists for three hundred years. As it turns out, a similar and unusual sixth toe, or "panda's thumb," serves the panda well as a means of grasping bamboo. Moles develop a sixth digit that aids in digging.[11] These sixth digits reverse the long trend of descent among tetrapods that stabilized at a count of five digits (after reaching eight in *Acanthostega*), and then continued as a frequent reduction from this number to four, three, two, or one load-bearing, grasping, or clawing digit. Front feet and hind feet may differ in the number of toes. For example, the crocodile has five on its front feet and four on its back. It is recommended not to get close enough to confirm this observation in the wild. Make this a zoo inquiry.

Performing elephants often bend down on their "front knees" (carpal bones) the same as do trained horses. Taking a bow "on bended wrist" would certainly sound peculiar. Knees can be found on the forelimbs of many four-footed animals, dogs and horses included. They have elbows higher up—in effect, "above the knee." There are therefore bones between their knees and elbows: the radius and ulna (mostly fused in the horse, these more twisty bones constitute the forearm in humans). Elephants, you see, do have knees—fore and aft.

So, what makes a knee a knee? Maybe a knee is a knee based on the geometry of how it bends. Maybe a knee is a knee because it forms the joint between the single femur and the paired radius and ulna, all capped by the patella. The answer depends on the namer: trainer or biologist. Naming is helpful to the animator, but the important task is to get the bones sequenced, proportioned, joined, and articulated correctly whatever their names.

As Chuck Jones realized, the hip of a sea lion is located near the end of the body, where the back flippers flare out. Of course, the human torso has a hip girdle at its lower end as well, but the legs put the human hip about midway between feet and head. In the sea lion, the hip and feet are in close proximity, the structure Jones achieved by binding two boys' legs together. People with legs bound can freely move only their feet, like a seal or sea lion.

A sea lion wearing trousers would need to tighten its belt just above its hip—at the waistline. Very short shorts would suit the sea lion just fine. Look closely at the flippers in order to find the fingernails. Or do you prefer to call them "flippernails"?

The zoo depends on recreational visitors, but zoo people expect to accomplish more than entertainment. Silly musings while limb-looking enhance engagement with the animals on display. Exhibit designs teach that habitat conservation is the best means of species preservation. Modern zoos are biodiversity arks that have inherited an entertainment and recreational role. This inheritance and the conservation message combine to craft visitor encounters that evoke appreciation for wildlife while aiding in the training of animators.

Zoos pay close attention to principles of artistry in crafting visitor encounters: exhibits are vivid depictions, arrangements of species cohere around themes, and multiple art forms focus attention. A charcoal sketch portrays a chimp's pensive facial expression; a bronze sculpture freezes a polar bear's efficient paddle; a series of panels isolates a penguin's displays of aggressive and submissive behaviors. Whether in verse or visual form, no exhibit is without art.

Sketchy Zoos

Sketching allows us to hold captive in the imagination the great diversity of limbed life. So distinctive are the feet and toes of hippos and storks, rhinos and bears, zebras and sea lions! Within these extreme differences deep similarities remain. The forelimbs of naked mole rats and fruit-eating bats end in fingers—clawed in one case, webbed in the other. The bone lengths vary, but the sequence of joints aligns from species to species, exactly as taught by Chuck Jones.

Pencil in hand, the amateur artist begins the zoo tour. Careful portraits of lumbering elephants and tottering penguins, posed in a series of positions, remain for now the professional animator's dream. Achieving realism, elegance, and complexity is not the goal: the line's the thing. The line segments depict angles and lengths. From shoulder to toe the segments zig then zag, some short, some long. The back may arch, run straight, or sag. With training, getting the proportions right becomes possible. A thumb moves along a pencil held in the line of sight; the segments marked by the thumb's position and the pencil's angle transfer to paper in correct proportion and orientation.[12]

Stick rhinos and stick hippos appear on the pages of the journal. The elephant's massive body reduces to skinny lines and angles. The necks of storks and giraffes seem almost silly in proportion to their bodies.

Trapezoids, triangles, squares, circles, ovals, and rectangles create heads, necks, limbs, haunches, shoulders, ears, muzzles, hooves, and tails. Soon a stocky mountain goat, then a bulbous Visayan warty pig, later a rotund orangutan, and eventually a long-limbed gibbon peer back from the pages.

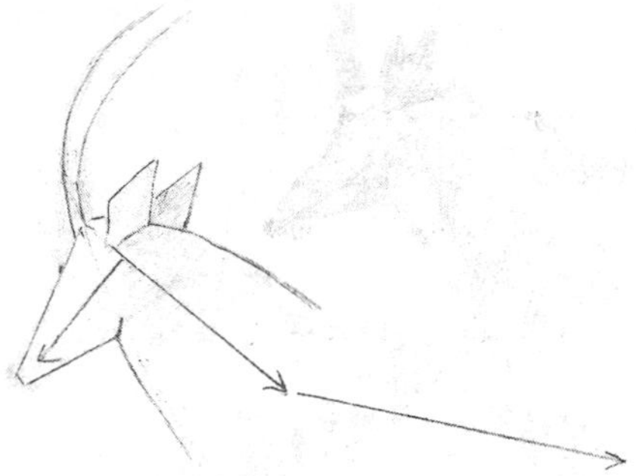

Figure 6.1 From angle to sketch of the neck and head. Illustration by Jan Glenn.

Figure 6.2 Angles, shapes, contouring, and detail in the sketch of an antelope's head. Illustration by Jan Glenn.

Two steps remain: (1) contouring the outline of the animal's profile and (2) finishing with shading. The process is Picasso in reverse: contouring the blocky impressions yields more realistic forms; then adding details of shading plus impressions of feathers or fur completes the picture. Ideally, a second sketch follows, completed in similar style yet focused on another feature of particular interest: the neck and head, with attention to dentition or feathers, for example.[13]

The ankle and elbow tour of the zoo guides attention to the proportions and shape of the body, too. By "going out on a limb" to its nailed, hoofed, padded, or clawed end, the artist hypothesizes the relative positions of knees, ankles, and elbows. Each change in angle implies a joint, shoulder to foot. The sketching mantra does the trick: make a stick figure, block in basic shapes, contour the edges, add some shading, finish with details to taste. Work quickly at each task. Be the artist you can most certainly be.

Make sure to fine-tune the digits. Do claws, hooves, or nails come in singles, doubles, triples, or more? Do they turn inward or outward or backward or align straight ahead? Which way do the hind toes point on creatures that climb: the sun bear, the margay cat, the local squirrel that chanced to run by? Note which parts of the foot or hand touch the ground when the creature walks and whether these contact points are the same or different when it stands or runs.

Attention to drawing details slows down the all-too-often frenetic tour of the zoo. You simply see more and very probably feel more. Style of movement, adaptation to habitat, and degree of common ancestry: all are bound up in limb structure. (The same could be said of skulls.) Awareness of the astonishing details of limbs, united by common descent and diversified by adaptation to different habitats, sows seeds of wonder.

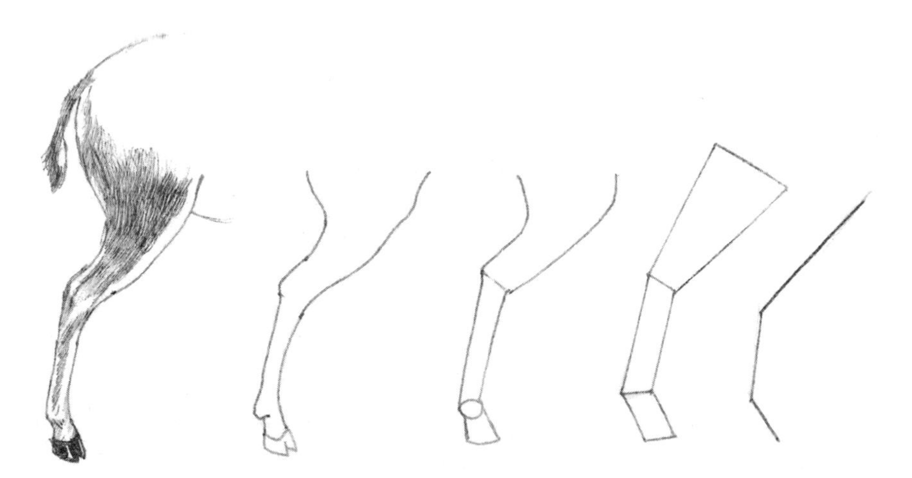

Figure 6.3 From line to limb. Illustration by Jan Glenn.

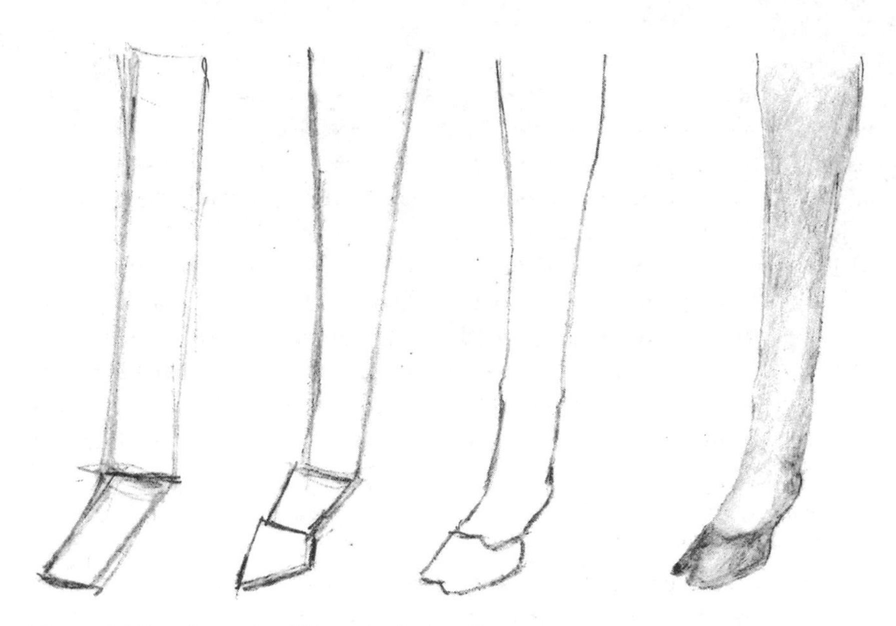

Figure 6.4 From line to hoof. Illustration by Jan Glenn.

Figure 6.5 Contoured image of antelopes at the zoo. Illustration by Jan Glenn.

Figure 6.6 Shaded image of antelopes at the zoo. Illustration by Jan Glenn.

Comparing limbs by means of sketches prompts archetypal Darwinian questions: "How closely related are bears and cats, dogs and bears, and dogs and cats?" You may make some good guesses based on the data acquired by sketching. Only cats, for example, can retract their claws. Claws are always visible on bears and dogs. One trait is not definitive, but the retractable claw does suggest that dogs and bears are the closer kin, with cats more distant on their tree of common ancestry. If looking closely at the zoo creatures turns you curious about these matters, some quick Googling will map out the affinities of mustelids (weasels, ferrets, martens, polecats, stoats, and more), pinnipeds (seals, sea lions, walruses), canines (dogs and wolves), ursids (bears), and felines (cats). At the zoo tour stage, the hand trains the eye and the eye stimulates the mind.

The Importance of Limbs

Darwinian histories portray the stunning creativity of natural selection.[14] The struggle for existence through time turned lobe-finned swimmers into walkers (tetrapods), walkers into runners, and then some back into flyers and swimmers. People, not nature, decide how to parse these groups. Crossing an imaginary boundary defines evidence of the first bird, *Archaeopteryx*, or an ambulatory whale,

 Ambulocetus. Limbs feature prominently in drawing these boundaries. At some point, however, the ancestor of the first bird was not truly a bird and the earliest progenitor of the whale was not an actual whale. Nevertheless, they had limbs that changed through tens of millions of years and left telltale signatures in the joints.

For example, the story of flighty *Archaeopteryx*'s wrist bones foreshadows that of scrambling *Ambulocetus*'s ankle bones. The former, a primitive bird, and the latter, a transitional whale, mark key steps in the derivation of species dramatically different from their ancestors. In both instances, limb bones trace the path of evolution.

Among a long-standing group of terrestrial mammals, at the end of the leg and start of the foot sits a short ankle bone, rounded at both ends like a double pulley: the astragalus. This bone defines the artiodactyls: antelope, goats, pigs, cows, sheep, deer (but not horses). These are even-toed, hoofed (and quasi-hoofed), toe-perching runners. Whales may be one of their offshoots, key evidence being that *Ambulocetus* had "double pulley" astragalus bones in its ankle joints. (The ends of the bone look like small paired wheels.)

Strange cousins emerge from a commitment to the Darwinian purpose of puzzling out evolutionary ancestry. For example, hardly anyone feels surprised to learn that whales are mammals. Within the mammals, whales represent a highly specialized and modified version of artiodactyl (even-toed) ungulates, the group of mammals that typically run swiftly on their toenails. Whales are so changed from the artiodactyl pattern that such an affinity is very elusive and difficult to recognize. Apparently whales and antelopes share an ancestor that lived tens of millions of years ago. That ancient creature's progeny turned into hippos, cows, deer, pigs, goats—and whales. Whales branched off sharply in an aquatic direction, shed their hooves, and earned a label of their own: cetaceans, a term derived from the Greek word for a giant monstrous fish. Cetologists claim that before evolving fully seagoing habits, very whale-ish creatures once walked and stalked ancient shorelines. Someday, somewhere—probably in Pakistan—a group of fourth-graders may find themselves picnicking next to a set of tracks left by a walking whale.

Classifying on the basis of limb anatomy ties whales to ungulates, tetrapods to lobe-fins, and birds to dinosaurs. Inferences about origins are embedded within classification. The comparison of wrist bones helps to solve the bird origins puzzle, similarities among ankle bones the whale one. Each story, told by comparative anatomy, amends the one, then two, then many bone limb pattern observed in *Tiktaalik* from 375 million years ago. Awareness of this stunning unity hiding amidst the variation of skeletal archetypes causes strange, enlightening, titillating, and eerie feelings.

The children of *Sesame Street* learn to classify by singing, "One of these things just doesn't belong."[15] Horses trot to a different song than cows, goats, and antelopes do: the horse "just doesn't belong." The ankles and toes tell why. The horse's hoof is a single toe, an ancient odd-number pattern, the same one that typified the three-toed horses of the fossil record. The same pattern tracks rhino fossils, from extinct gracile to modern robust forms. According to limb structure, horses and rhinos are quite closely related to each other. A good name for their common ancestor would be a "rhinorse" of course.

Is a rhino more closely related to an elephant or to a zebra? In terms of limb similarities, the rhino-zebra match is striking. Are giraffes related to zebras more so or less than they are related to storks? Relating storks to giraffes appears to be a stretch. In terms of hoofness and hipness, knee jerks and ankle twists, the stork is the one that just doesn't belong. Its wing is its thing.

Paleontologist Neil Shubin recommends touring the zoo and answering such questions as a means for understanding the history of vertebrate life. Observations of shared traits, says Shubin, "can be organized and arranged like a set of Russian nested dolls." Every vertebrate creature has a head and two eyes. One subset boasts limbs, and another, smaller one speaks and walks on two legs. That's the fish-to-us progression. For Shubin, this nesting also predicts where to look in the fossil record for potential ancestors—older and older in parallel to the nestings from the innermost subset to the outer ones.[16]

As you tour the zoo and relish the delightfully diverse ways that animals climb, swim, jump, dig, hop, run, fly, grasp, paddle, and crawl; take care to sketch joints in the same relative positions, body to toe, creature by creature. Think about an elephant kneeling down on its front legs, the risky behavior trunks make unnecessary (as told by Rudyard Kipling).

Sketching focuses on limbs, segment by segment, angle by angle. That's where Chuck Jones and his animators looked closely. Zoo sketching need not extend to animating, but Jones's primary insight is telling: structure determines how animals move. Movement, after all, is essential to survival. And Jones's boss, Walt Disney, intended realistic movement for his animated *White Seal*. The chase by the killer whale had to excite the audience with convincing realism in a cartoon format. Baby seals, the little ones who must not swim until they are six weeks old—lest, as Kipling wrote, their heels sink their heads—are indeed Disney-cute. They need time to develop their flipper muscles. Flippered swimming, the pinniped way, may lack the power and grace that flukes give to whales, but it offers dramatic maneuverability. Seals can indeed snatch fish with great success.

Chuck Jones bound the hind limbs of his grandsons together, hip to ankle, and tied their forelimbs at the elbow to their bodies. Tossed into the pool with swim fins on hands and feet, the boys mimicked the swimming motions Jones's animators were to re-create. He had forced their limbs to articulate in the same relative positions he had observed among seals and sea lions. The modified structure of their limbs determined their movements. Jones's autobiography makes no mention of whether he next tossed a few goldfish into the water for his grandsons to chase. If he did, the goldfish had little to fear.

7

NOSEY ELEPHANTS

A Tale of Trunks and Tusks

"Come hither, Little One," said the Crocodile, "for I am the Crocodile," and he wept crocodile-tears to show it was quite true.

Then the Elephant's Child put his head down close to the Crocodile's musky, tusky mouth, and the Crocodile caught him by his little nose, which up to that very week, day, hour, and minute had been no bigger than a boot, though much more useful. . . .

Then the Elephant's Child sat back on his little haunches, and pulled, and pulled, and pulled, and his nose began to stretch. And the Crocodile floundered into the water, making it all creamy with great sweeps of his tail, and he pulled, and pulled, and pulled.

And the Elephant's Child's nose kept on stretching; and the Elephant's Child spread all his little four legs and pulled, and pulled, and pulled, and his nose kept on stretching; and the Crocodile threshed his tail like an oar, and he pulled, and pulled, and pulled, and at each pull the Elephant's Child's nose grew long and longer—and it hurt him hijjus!

Rudyard Kipling, "The Elephant's Child"

Full of 'satiable curtiosity and asking ever so many questions, the Elephant's Child had an unquenchable appetite for learning. Numerous spankings received for asking impertinent questions failed to dim his curiosity. The Elephant's Child wanted only to know what the crocodile had for dinner and the Kolokolo Bird sent him to the great grey-green, greasy Limpopo River to find out: "My father has spanked me, and my mother has spanked me; all my aunts and uncles have spanked me for my 'satiable curtiosity; and still I want to know what the Crocodile has for dinner!"[1]

No doubt generations of children, whether they have seen the banks of the great grey-green Limpopo River or not, have learned from reading *Just So Stories* to be wary of muddy waters. The tale of the Elephant's Child does teach a lesson in safety with gentle humor. At the same time, the story tells how the young elephant had to struggle with his question. The answer came not only with difficulty but also with a surprise.

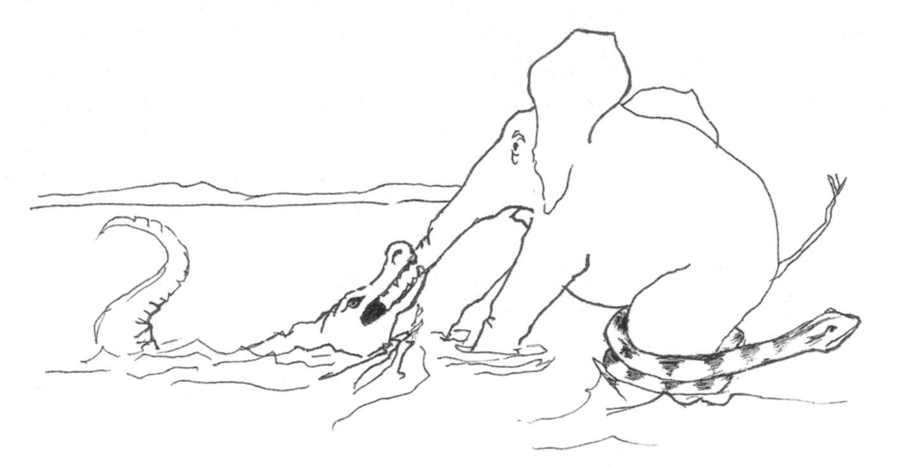

Figure 7.1 The crocodile schools the Elephant's Child. Illustration by Jan Glenn.

Limpopo crocodiles will eat a baby elephant—if they can. Failing in that endeavor, they may stretch the bulgy boot of a nose into a trunk, an adaptation that, according to Kipling, quickly spread among the elephant population as "a really truly trunk same as all Elephants have to-day." No longer would the Elephant's Child need to walk about "eating melons, and throwing the rind about, because he could not pick it up." He found he could pick up melon rinds with his trunk and had quite fortuitously acquired the skills of fly-swatting and snorkeling.

> So the Elephant's Child went home across Africa frisking and whisking his trunk. When he wanted fruit to eat he pulled fruit down from a tree, instead of waiting for it to fall as he used to do. When he wanted grass he plucked grass up from the ground, instead of going on his knees as he used to do. When the flies bit him he broke off the branch of a tree and used it as fly-whisk; and he made himself a new, cool, slushy-squshy mud-cap whenever the sun was hot. When he felt lonely walking through Africa he sang to himself down his trunk, and the noise was louder than several brass bands.[2]

Indeed, elephants do pluck grass, pick fruit, trumpet messages, whisk flies, and roll mud balls with their trunks. For Rudyard Kipling's young fans, the Elephant's Child's encounter with a crocodile whimsically explained how the elephant got its trunk. Once home, the Elephant's Child displayed his trunk's prowess and all the other elephants just had to have one. Trunks spread throughout the population as everyone hurried "to the banks of the great grey-green, greasy Limpopo River, all set about with fever-trees, to borrow new noses from the Crocodile." To this very day, "O Best Beloved, all the Elephants you will ever see, besides all those that you won't, have trunks precisely like the trunk of the 'satiable Elephant's Child."[3]

The Elephant's Child's inquisitiveness suffered a bit from the hunger of a predatory beast. His questioning engaged him in adventure, risk, and struggle. The Elephant's Child returned to society to share the wonders—bugling, fly-whisking, and grass-plucking—so serendipitously acquired. That's the essence of the Kipling story.

Kipling succeeds in teaching basic natural history about elephant adaptations—what the biology teacher would refer to as structure (trunk) and function (fly-whisking, snorkeling). Children learn as well to watch out for crocodiles. The question of origins remains. How did trunks, and for that matter elephants, come to be? For scientists, the same as for the Elephant's Child, the answer comes not only with difficulty but also with a few surprises.

What Is a Trunk?

Some time ago, the noted anthropologist and philosopher (and husband to Margaret Mead) Gregory Bateson pondered the question "What is an elephant's trunk?" He emphasized that a nose is a nose by context: by its connection to the elephant body pattern. Argued Bateson, "That which stands between two eyes and north of a mouth is a 'nose,' and that is that."[4]

During its embryological growth, a "bulgy boot" of tissue and cells responds to messages from other tissues by growing and differentiating, as if pulled by a crocodile's bite. In fact, the trunk is something more than a tubular, flexible nose. Look closely; beneath the trunk there is no upper lip. In one sense, a trunk is a long, skinny, and definitely not-very-stiff upper lip with nostrils.

Kipling's crocodile pulled on the Elephant's Child's upper lip as much as on its bulgy nose. The upper lip is also north of the mouth and generally south of the nose. Lips are finely muscled in animals that graze or browse, and attachment to the elephant's upper lip, in effect, makes the nose mobile. Lips and noses fit nicely together on the face of an elephant and add reach to its mouth.

So, an elephant's trunk is more than just a nose. It's good for doing many things useful to elephants, as Kipling explained. Trunks develop from non-trunks in two senses: first, within elephant embryos, and second, over time, from elephant ancestors. Kipling's crocodile works well as a metaphor for the pull of natural selection working to fashion trunk-ish novelty.

Trunks lift logs as easily as coins. Through openings in the elephant's skull pass bundles of nerves and vessels of blood that innervate and power the trunk. Boneless muscle spans its length. The trunk ends in a sensitive lip as dexterous as a finger. The trunk may caress as easily as it may strike with violence. It sniffs and snorkels and trumpets and whisks. Elephants even test unfamiliar ground with their trunks to make certain the footing is safe.

There is no evidence that early proboscideans ("proboscis" means flexible snout) had trunks,[5] though they likely did live among crocodiles. According to Kipling's story, trunk-getting worked in reverse of Darwin's descent-with-modification approach. The child passed its trunk along to its parents, a biologically improbable event. In all likelihood, elephants of yore bequeathed trunks to elephants of today without a crocodilian middleman.

The Story Fayum Fossils Tell

Often the use of disarmingly childlike logic helps in thinking Darwinian thoughts. Consider inheritable traits such as skeletal structure. The greater the similarity, the more recently the organisms under consideration likely shared a common ancestor. It follows from this method that if great similarity implies close relationship, great dissimilarity implies distant branching. This Darwinian logic, set to the tune of the familiar *Sesame Street* song, "One of These Things" becomes "One of these bones is not like the others. One of these bones just doesn't belong." The ones most alike are from the more closely related creatures. The others are from more distant family members.

For example, Asian and African elephants are similar to each other, but the ice age woolly mammoth is more closely related to the Indian elephant than to the African one: the African one "just doesn't belong" to the tight Asian woolly group. The bones of these three species, in turn, are more similar to one another than they are to those of the extinct mastodons. Mastodons in this sense are the ones "not like the others."

Most likely, nubby-nosed creatures wading among crocodiles tens of millions of years ago were ancestral to them all. Kipling's tale intersects with paleontology's own story, for there actually are nubby-nosed proboscids in the fossil record: the extinct moeritheres. These beasts were modern elephants' ancient relatives

The story begins with digging for fossils in the Fayum district of Egypt and the discovery of bones from a pig-sized animal with enlarged pairs of upper and lower incisors, "the beginning of tusk formation . . . [and] the first faint beginning of the elephant tooth pattern" of parallel, tightly compressed cross-ridges.[6] This creature, *Moeritherium*, had no trunk, but perhaps had a flexible, bulgy boot of a snout. Its fossil bones lie in beds dated at approximately 40 million years of age. Beds a few million years younger contain preserved bones of *Palaeomastodon*—a creature easily recognized as an elephant's not terribly distant cousin. Entombed in Egypt are more than mummies!

Fossil remains from Morocco of a creature that lived approximately 60 million years ago hint at an even earlier origin of proboscid form. The creature, *Eritherium*, weighed in at just a few kilograms—the size of a big bunny.[7] Evolutionary biologists have concluded that these ancient proboscids diverged from common ancestry with other African lineages early on in the history of placental mammals. As a result, today's elephants claim the sirenians (manatees and dugongs, sometimes referred to as "sea cows") as their closest living relatives. Molecular evidence places elephants in close relationship with the sea cows, followed by affinities to the hyraxes, aardvarks, elephant shrews, tenrecs, and golden moles.[8] The Elephant's Child could indeed boast quite an assemblage of distantly related aunts, uncles, and cousins across Africa.

Skull, wrist, and tooth anatomy place *Moeritherium* in the morphological elephant ballpark, a trunkless yet bona fide member of the proboscid clan. Whether or not modern elephants descended directly from the Fayum's bulgy-nosed, swamp-strutting creatures of 35 or 40 million years ago cannot be known with any certainty. From the cusp patterns of their teeth to the joint structures of their front feet, moeritheres represent the primitive state from which elephants might readily be derived. *Moeritherium* ambled its way hippo-like through mangrove forests, rivers, and marshes of the Late Eocene epoch, earning its living as neither a wholly aquatic nor a wholly terrestrial creature.

The marshland habits of the ancient proboscid moeritheres and the snorkeling function of the modern trunk seem to suggest elephant origin in an aquatic habitat. Proboscids, however, have clearly enjoyed relatively arid lands for much of their history. How convincing is the evidence for an out-of-the-water claim for elephant origins? Those who try to answer such questions combine anatomical, embryological, and molecular data with fossil records. An example of "just how" follows.

Arguing Descent from Descent

A provocative article about elephant origins appeared at the end of the last century: "The Developing Renal, Reproductive, and Respiratory Systems of the African Elephant Suggest an Aquatic Ancestry."[9] Zoologists Ann P. Gaeth, Roger V. Short, and Marilyn B. Renfree claimed that developing elephant embryos and fetuses display features expected among aquatic organisms but rare or absent among terrestrial ones.

Looking for clues of ancestry in patterns of embryological development has a long history in evolutionary biology. The research by Gaeth and her team illuminates this approach. Of course, the best science happens in the context of debate and more recent research has refuted much of their argument for aquatic elephant ancestry.

The authors of the "Aquatic Ancestry" article characterized the fetal development of the elephant's trunk as "precocious," meaning the trunk appears early in development and is well formed at birth. To them that suggested a feature with a long pedigree. Trunks remain unknown in the fossil record of early proboscids, however. The embryological clue by itself is quite speculative.

Gaeth and her colleagues also noted that elephant testes do not descend, the same pattern observed among present-day seals and whales. Thus, their study of descent crosses from the description of fetal development to comparative anatomy across species—two kinds of evidence bearing upon an origins hypothesis. Could the anatomical story of elephant testes mean a watery origin for proboscids? In combination with the development of the fetal trunk, it entices such a conclusion.

Alas, the testicular evidence appears to lead to a dead end. Most interestingly, both sides of the debate about the significance of this trait for inferring origins make comparisons across living lineages. The Gaeth report stresses retention of testes in the abdomen among marine mammals. The critics find more convincing evidence in molecular data linking elephants to all of the other Afrotheria: "Lots of land mammals lack external testes. Among them specifically are (but not exclusively) all of the Afrotheria, a large group of mammals with African origin and the major subgroup of mammals to which elephants belong, including hyraxes, aardvarks, and elephant shrews."[10]

Testes also do not descend in the monotremes, a primitive group of egg-laying mammals (now represented by the platypus and echidna). In summary, according to the molecular evidence, Afrotheria dispersed and differentiated rather early in the history of placental mammals, carrying their testes internally.

Obviously, other groups of placental mammals evolved testicular descent. Science historians Karel Kleisner and Jaroslav Flegr, together with molecular biologist Richard Ivell, examined the diversity of mammalian testicular positioning (held within the abdomen or carried externally in a scrotal sac). In effect, they examined "descent" in order to evaluate descent, and suggested "that the recent diversity in testicular position within mammals is the result of multiple selection pressures stemming from the need to provide conditions suitable for sperm development and storage, or to protect the male gonads from excessive physical and physiological disturbance."[11] By disturbance they meant behaviors such as galloping.

Keeping the testes ensconced within the abdominal wall was probably the "original ancestral state" of placental mammals and maybe even of all mammals, argue Kleisner, Ivell, and Flegr. It's certainly the afrothere way. Their analysis of anatomical evidence discounts the evolutionary loss of external testes in a scrotum. In the primitive state the testes are held close to the kidneys in the dorsal (top) side of the abdominal wall. They argue that returning testes from the scrotum to the abdomen would place them ventrally (bottom side). Elephant testes are right where one might expect them to be in the "original" state—the top side of the elephant abdomen. If through the course of evolution the elephant's body has reabsorbed them from a descended condition, then they are on the wrong side of the kidneys.

Noted vertebrate paleontologist Christine Janis draws the conclusion succinctly: "Elephants and other afrotheres appear to be primarily testicond (undescended testes) whereas seals and whales are secondarily so. The point being that, if descended from a group that had a scrotum originally, internal testes could be an indication of aquatic tendencies, but not in a group primarily without a scrotum."[12] Proboscids never had a scrotum to lose. Therefore its absence indicates nothing about a watery origin. Whales and seals are not afrotheres; their lineages likely did reverse the path of testicular descent—making them "secondarily testicond," in lovely technical jargon.

Funnel-Shaped Kidney Ducts

Gaeth, Short, and Renfree have one more embryological argument up their collective sleeves, as alluded to in the title of their article. They have found a ghostly clue linked to kidney formation. At issue are minute structures that form early in elephant development and are lost by birth: "nephrostomes." All elephant fetuses have these funnel-shaped kidney ducts. They function in flushing waste out through the abdominal lining and in exchanging water and salt between blood and body fluid.

Until Gaeth's team meticulously examined fetal elephant tissue, no one had ever observed nephrostomes in live-birthing mammals. Not all mammals birth live babies: monotremes, the echidna and platypus, lay eggs. Nephrostomes are found in monotremes but, until witnessed in tiny elephant embryos (weighing about half a gram), never before among the viviparous mammals.[13]

What drove Gaeth, Short, and Renfree's support for the aquatic origins hypothesis was the commonness of nephrostomes among freshwater vertebrates from frogs to sturgeons. As vertebrates evolved terrestrial lifestyles, they took the aquatic habitat with them as egg sac fluid. Egg-laying reptiles and all birds, they pointed out, develop nephrostomes in their earliest developmental stages. Elephants, perhaps having retained a primitive feature discarded by placental mammals, seem a throwback to egg-laying times or an adaptation to watery environs.

Or perhaps not. Elephants have a very long period of gestation, and fetal growth is slow. Rudimentary nephrostomes blink in and out of existence in the very first embryonic stages of some mammals (domestic cat, common bushtail, sheep). Maybe their appearance in elephant embryos is no more than a side effect of slow growth, though rhinos and whales, other big mammals that experience relatively long gestation periods (but still shorter than elephants'), have no fetal nephrostomes.[14]

Deep Snorkeling

Keep in mind that the elephant's minuscule embryonic trunk is destined to grow into an eight-foot-long snorkel. That's the singular trait that so captivated Rudyard Kipling and begged for an explanation. Roger Short, the reproductive biologist on the Gaeth team who obtained the collection of embryos for research from elephant carcasses, found another trait to be quite fascinating: the attachment of the lungs to the chest wall. Elephants lack a pleural cavity.

"Why doesn't the elephant have a pleural space?"[15] One answer: the better to snorkel with at depth. Collagen fiber fills this space connecting the lungs to the wall of the chest.[16] To Short, this situation made excellent snorkeling sense. Deep snorkeling elephant-style is very dangerous, as explained to Short by a respiratory physiologist:

> If you're a snorkeler, you know that you're not allowed to have a snorkel tube that's much longer than [about sixteen inches] because, if you do, you will actually rupture the blood vessels in your chest cavity. . . . And here is an elephant with a snorkel tube that is about eight foot long so they couldn't possibly snorkel, were it not for the fact that they have managed to glue their lungs to the chest wall so that they can't get a haemothorax.[17]

In brief, a long snorkel for filling lungs "glued" to the chest wall enables the swimming talents of elephants. The lung arrangement overcomes the danger of blood leaking into the pleural cavity and restricting the lungs' ability to expand.

Prehistoric elephants crossed the gulf from the California mainland to Santa Catalina Island, and fishing trawlers have snared mammoth tusks, teeth, and vertebrae from North Sea sands.[18] The snorkeling trunk appears to have been essential to elephant dispersal. The question remains: Was the snorkel retained from an aquatic ancestry, or did having a trunk and a collagen-filled pleural cavity—evolved as an adaptation for terrestrial life in arid and grassy habitats—serendipitously create a sea-crossing opportunity for elephants?

There's just no evidence of trunks from the fossil record of early proboscideans. And the gluing of lungs to chest walls has a non-snorkeling explanation: it's a way for big bulk to adapt to gravity, a mechanism that simply differs from what works in smaller mammals. The downward pull of an elephant's weight tends to compress the lungs. An elephant simply could not survive with the pleural cavity arrangement common to less hefty creatures. It might easily suffocate itself.[19]

The study of embryos and the comparative analysis of anatomical and physiological traits reside at the crossroads of debate about the origins of species—their Darwinian histories. The fossil record, of course, despite its incompleteness, plays an essential role. It features bones and teeth. One recent analysis, for example, of the oxygen isotopes in moerithere tooth enamel has yielded matches to the patterns common to aquatic and semiaquatic mammals that live on a diet of aquatic vegetation. The enamel composition found among mammals from these habitats significantly differs from that of terrestrial ones, and this line of evidence suggests an aquatic lifestyle among early un-trunked proboscideans.[20] It says very little, however, about how the Elephant's Child and his ilk got their trunks.

Comparative studies of DNA often provide the most convincing evidence of relative common ancestry. The ability to analyze DNA to determine, in effect, that a finding of "one of these sequences is not like the other" often settles many questions about the relative recency of a common ancestor. Although the fossil record is rather stingy with regard to DNA and rather generous with respect to hard body parts, the comparative method works analogously whether in reference to DNA or to skeletons. For example, genetic sequencing underscores the common afrotherian ancestry of hyraxes (conies), proboscideans (elephants), and sirenians (manatees and dugongs).[21]

Selecting Proboscids

"What caused new species to appear in place of the extinct ones?" wondered Darwin and his contemporaries. Today's African elephants must fend their way across Africa amidst a landscape of crocodiles, flies, melons, mud, great grey-green, greasy rivers, Kipling's Bi-Coloured-Python-Rock-Snakes, and much more. These are the conditions of life exploited by individual variations among noses and every other structure, all performing in concert to promote survival.

Darwin calculated that if elephants entered reproductive maturity at age thirty and continued to breed until age ninety, producing six offspring during this time and living until one hundred years old, then within "750 years there would be nineteen million elephants alive descended from the first pair"—quite a few elephants' children.[22] Only a tiny fraction of this potential number of elephants ever live; they compete with one another, vary in minute respects from one another, and differ in their reproductive success. Many die young.

During the course of the last several tens of millions of years, hundreds of proboscidean species have roamed Europe, the Americas, and Asia; the Asian and African elephants struggle to survive today. The evolving thicket of their ancestral relatives exploited tropical swamps and glacial margins. Why have so many gone extinct and only two, the Asian and the African, remain? Perhaps populations adapted narrowly to very specific habitats, thus becoming vulnerable as climate changed. Perhaps human hunters dispatched tottering populations of woolly elephants as the ice age waned. The two surviving species do exhibit flexibility, for they can range across a variety of landscapes, consuming diverse foods.

Competition favored woolly coats when climates turned cold. These elephants suffered an extinction witnessed (and perhaps encouraged) by our own ancestors. As Darwin wrote in the *Origin*: "The fact of the extinct elephant and rhinoceros having formerly endured a glacial climate, whereas the living species are now all tropical or subtropical in their habits, ought not to be looked at as anomalies, but as examples of a very common flexibility of constitution, brought, under peculiar circumstances, into action."[23] In other words, variation under selective pressure brought about bare-skinned elephants, well suited to life in warm climes. The trunk, as a food-gathering instrument, contributed to their flexibility.

"This preservation of favourable individual differences and variations, and the destruction of those which are injurious, I have called Natural Selection, or Survival of the Fittest," famously wrote Darwin. The struggle to exist within variable habitats in concert with variation among individuals promoted what Darwin termed "diversification of structure." Diversification of structure depended upon the heritability of variations. "Unless favourable variations be inherited by some at least of the offspring, nothing can be effected by natural selection."[24]

Descendants differ, but do not depart entirely, from the forms of their ancestors. No offspring, unless cloned, duplicates the parent exactly. New traits are not written on blank slates. They are derived from old parts proving useful in novel ways under new and unexpected conditions.

Natural selection is a blind opportunist. It converts unplanned variation into new creatures changed in surprising ways from their forebears. At the same time, inheritance embeds in the skeletons, organs, and molecules of each generation the signatures of their ancient progenitors.

And "just so," by means of natural selection, the nubby boot became a trunk.

Unfortunately, the phrase "just so" masks the interesting details of "just how?" Selection pressure apparently yanked a trunk out of a bulgy boot of a nose by means of its success—as a snorkel, flyswatter, dust-spreader, communication apparatus, and perhaps most important, an implement for feeding and drinking. But elephants were on their way to becoming elephants before trunks took form. Assuming the crocodile story to be an allegorical version of how natural selection alters a large herbivore's mouth parts, just how might the Elephant's Child have obtained a trunk? A plausible answer casts several actors in the role of the crocodile: habitat, mating, and head variation.

Big Bites and Natural Selection

Elephants, and other large herbivores such as giraffes and rhinoceroses, must forage over wide areas to find the food they need to sustain themselves. The teeth, jaws, tongues, and lips of these large herbivores must work in concert to maintain the intake their body mass demands. Otherwise they starve. A big head (not to mention a big gulp) may help—or at least big lips or a very big tongue, coupled with teeth able to process ample intake.

Proboscids have endured for tens of millions of years, escaping extinction events that doomed many of their fellow super-sized browsers and grazers. The end of the Eocene (about 34 million years ago) witnessed cooling and drying, leading to shifts in vegetation. Massive dual-horned, rhino-related brontotheres, for example, struggled to meet the requirements of their specialized browsing habits—a possible reason for their extinction as the subsequent Oligocene world opened.[25] Moeritheres, the elephant relatives, had thrived during the Eocene but, according to their known fossils, apparently failed to survive into the Oligocene as well.

The changing climate of the Late Eocene brought increased seasonal variation and changing floras with consequences for herbivore evolution. Plants likely responded to climate change with adaptations that would impact feeding habits. For example, by differentiating the fiber content of leaf and stem, the evolving flora would have created opportunities "for the selective feeding habits that characterize present-day ruminants."[26] Climate changed, plants changed, plant-eaters changed: an evolutionary trifecta.

Large herbivore guts need to detoxify plant toxins, and the specialized teeth of grazers must resist abrasion. Ruminating herbivores digest plant matter (including cellulose) in their guts by fermentation. Elephants and rhinos, for example, do so in their hindguts, cows and deer in their foreguts. The hindgut fermenters attain the greatest bulk—and correspondingly, require a mouth able to ingest copious amounts of food.[27] Grass, for instance.

Grasslands first took hold in the higher latitudes of the landscapes of 25 million years ago, with the evolution of abrasive-resistant teeth in close pursuit. Grazing specialists first took the stage about 10 million years ago. Miocene proboscid species flourished, "reaching their zenith during the middle to late Miocene."[28] Nearly all true grazers are even more recent (Pliocene and Pleistocene origins—the last few million years), a response to the spread of grassland and savanna habitats that differed in timing from one continent to another as the world felt the effects of a cooling (and increasingly volatile) climate. The complex cheek teeth of true elephants—as distinguished from other families of proboscids in the fossil record—express a grazing lifestyle suited to modern savannas, a grassland habitat only a couple of millions of years old.[29]

Mouths and guts adapt or species perish. Seasonal variation in the Eocene, the origin of grasses, Oligocene habitats varying from woodland to brushland, grasslands expanding in the Miocene, savannas spreading through Africa in the Pleistocene: each change challenged herbivore populations. Climate, not a crocodile, likely elongated the elephant's upper lip. Evolving a trunk may have required millions of years, but with it, along with the proper teeth and tongue, an elephant's mouth copes quite well with the distribution and type of food found in its modern environment.

Mouths unable to cope with shifts in nature's abundance doomed some creatures, while variations trending toward abrasion-resistant teeth and elongated, flexible upper lips created opportunities for others. That's a natural selection account based on feeding and foraging adapting to environmental changes rather than elongation by means of crocodilian predation. In Darwin's words:

> We shall best understand the probable course of natural selection by taking the case of a country undergoing some physical change, for instance, of climate . . . and some species might become extinct. . . . In such case, every slight modification, which in the course of ages chanced to arise, and which in any way favoured the individuals of any of the species, by better adapting them to their altered conditions, would tend to be preserved; and natural selection would thus have free scope for the work of improvement.[30]

Certainly among Darwin's key insights was that any "slight modification" favoring an individual's survival as conditions altered arose by chance.[31] Unequivocally Darwin concluded, "Unless profitable variations do occur, natural selection can do nothing."[32] Changes in vegetation do not cause elephants to birth children with elongated upper lips any more than tug-of-war with crocodiles creates inheritable trunks. Mouth parts vary fortuitously, not on purpose, in ways that keep food coming and likely enhance survival in other ways as well. Individuals vary; populations adapt as reproduction propagates fortuitously favorable variations.

Darwin furthermore believed that a slight advantage—"still further modifications of the same kind"—might set in motion a positive feedback loop: "Extremely slight modifications in the structure or habits of one inhabitant would often give it an advantage over others; and still further modifications of the same kind would often still further increase the advantage."[33]

South African zoologist Yolanda Pretorius and her coauthors reason that "enlarged soft mouth parts not only allow larger animals to be more selective but also to cover a bigger area in one bite."[34] In the Darwinian story, creatures able to swipe great bites of vegetation with efficiency gain a competitive edge. Specialized mouth parts enable large herbivores to eat not just more but also more nutritious vegetation because soft, dexterous lip, tongue, and trunk muscle may strip away the least nutritious plant material before swallowing.

The reach of the lips in elephants and rhinos and the extension of the tongue in giraffes extend the gape of the mouth. These are the keys to a big bite. From detailed field observations of foraging behavior, Pretorius and her colleagues conclude: "Megaherbivores of today such as giraffe, white rhinoceros and elephant have extreme mouth morphologies and sizes as a result of natural selection to adapt to the spatial mass and quality distribution patterns of plants."[35]

In summary, elephants have trunks because great body size requires mouth parts that forage with efficiency, especially when food is widely scattered and of varying quality. A trunk is sort of an outside-the-mouth tongue that works like a sweep net. One answer to how the elephant got its trunk is "diet did it." But it takes more than a trunk to make an elephant.

What trunks may gather, teeth must chew and throats must swallow. The tongue maneuvers to make these processes happen. Its jawbone attachment is just part of what helps it move around. A complex of muscles, ligaments, and small bones—the hyoid apparatus—also helps to control its motion. Among the proboscideans, for nearly 25 million years this structure, composed of five delicate bones, has had a distinctive, flexible form—less boxy than in other mammals. Swallowing depends on action of the hyoid. The structure of the proboscid hyoid apparatus allows for the lowering of the larynx in the throat to form a pouch behind the tongue and the flexibility needed to produce low frequency sounds useful among elephants for communicating across long distances. The pouch resonates to amplify vocalizations and when empty functions to store water.[36]

Other animals typically have several more hyoid bones than the elephant. The muscles, tendons, and ligaments that occupy the space left by the missing bones contribute to the formation of the pouch. Calls made at frequencies resonating between 5 and 24 hertz (infrasound) are inaudible to humans but detected by the specialized structures of the elephant's ear at distances of up to 2.5 miles. Infrasound propagates through both air and ground and thus elephants have the ability to detect seismic vibrations. Elephants not only make a variety of calls using their throats. They can also beat their trunk on hard ground or a tree to send an acoustic signal. Even a slap on the tusk with the trunk works quite well—one more trick for the Elephant's Child to teach his kinfolk![37]

The flexible hyoid structure is a distinctive feature of elephants. Its origin appears to coincide with arid conditions advancing across changing landscapes at the time marking the boundary of the Oligocene and Miocene epochs. As highly social elephant herds ranged widely to secure dispersed food, infrasound aided communication, self-showering cooled hot bodies, and pouched water quenched thirst. So goes the natural selection version of the story in terms of the elephant's hyoid apparatus—a trait hidden from view but equally diagnostic, if not more so, of elephantness as the trunk.[38]

A twenty-first-century Kipling might author a new just-so story: "How the Elephant Got Its Hyoid Apparatus":

> One brutally, blistery, very hot day the Elephant's Child knelt for a drink just as the bony Hyoid Fish came swimming happily by. The Elephant's Child swallowed that bony little fish. It stuck in his throat, O Best Beloved, but it didn't choke him. The Hyoid Fish struggled to escape and in so doing stretched the Elephant's Child's throat. He tried coughing to get rid of it but each time he coughed, it went deeper and deeper—and so did his voice. Soon he was speaking in mumbly, rumbly whispers that only

other elephants could hear with their big, wide heads. Soon all his aunts and uncles wanted to speak in such low voices too. And one by one they went to the river hoping to swallow a bony Hyoid Fish. And that, O Best Beloved, is how elephants came to be able to talk in rumbles to each other when far, far apart.

Combat and Sexual Selection

From an evolutionary perspective, finding a mate has great impact. Failure carries a heavy price. Selecting a mate and competing for mates provide remarkable leverage over behaviors and traits among both males and females. Frog song, bird plumage, goat head-butting, and beetle horns stand out among the results. Such "sexual selection" very likely propelled the evolution of tusks among some groups of proboscideans. The modern African elephant, *Loxodonta africanus*, is an excellent example.

Elephant tusks are greatly enlarged, gently spiraling incisor teeth. Their growth demands substantial allocation of energy and mineral resources as well as proper skeletal and musculature support. Tusks are quite expensive to grow. Elephants, especially the bulls, would not invest in them without the promise of a profitable return.

Tusks can do the traditional work of teeth but on a very grand scale: breaking up food. In effect, smashing tree bark and digging at salt licks extend the elephant's bite. Natural selection seizes whatever opportunity comes its way to enhance the prospects of survival. Tooth selection often results, of course, because successful eating is crucial for all creatures. Natural selection balances better teeth for better eating with their mineral and energy price tag. Sexual selection tips the balance in favor of much higher payments in return for extremely enlarged combat ready teeth needed by bulls to solve their most difficult reproductive challenge: access to females.

The problem the bulls must confront is not a lack of females but the lack of female receptivity to fertilization (estrus): a bit less than one week every four years, assuming two years of pregnancy and two years of nursing. This basic fact of life creates an intensely competitive situation: "Victorious males wield the longest tusks and tower over their rivals, standing more than twice the height of smaller males. To the victor go the spoils and, in elephants, this means the oldest, largest, best-armed bulls sire the offspring."[39]

Tiny sperm depend on giant tusks triumphing in combat for the chance to fertilize the egg. This situation has prevailed for a very long time among elephant ancestors and appears to have influenced multiple styles of tusk development. Some species, now extinct, sported greatly enlarged lower incisors (deinotheres), and even both upper and lower ones (gomphotheres). Among the proboscideans, this fearsome foursome of tusks predated the later condition of mastodons and elephantids: two enormous upper incisor tusks, a shortened lower jaw, and a broad, domed head. Evolution dispensed with the lower tusks.

Anancus, a mammoth of lesser size, reached just ten feet in height yet carried tusks thirteen feet in length.[40] It takes a very big head held very high to make room below it for such very big tusks. Muscle and bone must be proportioned accordingly. And then there's the trunk that must reach around and beyond the tusks to transport food to the elevated mouth.

Weaponized teeth are not only expensive to produce but also costly to use, in terms of injury and effort.[41] The mating payoff must indeed be substantial, a calculation performed by sexual selection across eons—a key to the origin of animal weaponry extremes, whether elephant tusks, deer antlers, crab claws, or dung beetle horns, argues Douglas Emlen in *Animal Weapons*.

Maybe the trunk became elongated in tandem with the enlargement of the tusks. Both work best with an elevated head. And having the ears set in a big, wide head helps to improve the audibility of low frequency sounds[42] There would seem to be little evolutionary advantage in plowing the ground with tusks with each and every step and dragging the trunk in the dirt, always having to be careful not to trip on it. Which came first—the trunk or the tusk—and the extent to which one enabled the other is difficult to tell. The elongating trunk kept the elephant's nose and part of its mouth close to the ground as its size increased and as its head became even bigger in relation to its body.

Early proboscids from more than 40 million years ago lacked trunks but may have had quite flexible upper lips. By 24 million years ago they had evolved hippo-like tusks from their incisors. Miocene time witnessed a dramatic display of proboscidean tusks: "Miocene tusks greatly increased in length and diameter; some reached over three meters in length and over 20 cm in diameter. They are the largest known teeth of animals, living or extinct."[43]

Through time, tooth count decreased, tooth size grew, and cheek teeth specialized. Proboscid teeth, along with the head, evolved more rapidly than other organ systems.[44] In other words, as the body got bigger and climate shifted, the elephant's gut changed less than its head. The engine ran well, provided it remained well fueled. Variations among lips, nose, teeth, and throat proved successful at exploiting the changing landscape.

Meanwhile, lengthy gestation and two years of nursing kept females out of the mating game for extensive periods of time. The intense competition among males for the chance to mate placed a premium on combat readiness. Sexual selection seized serendipitously on the evolving organs of the head and their propensity to evolve relatively rapidly. As a result, great masses of ivory came to adorn many proboscid species, the African bull bush elephant being the most striking modern example.

And the trunk came too. Ironically, its numerous functions, from snorkeling to spraying, included greetings. Elephants have no hands, but they do practice "trunk-shakes." Upon meeting, elephants often use their trunks to reach out to touch each other's face. Sometimes they intertwine their trunks, perhaps giving each other a measure of assurance.[45]

Finally, as the saying goes, "An elephant never forgets." Elephants' big heads house big brains, relatively large even for their body size. That gives them the capacity to remember where to look for food and water during droughts,[46] as well as an ability to use and make rudimentary tools, such as flyswatters.[47] Lifelong bonds and complex hierarchies suggest ample social intelligence as well.[48] After all, "an elephant's faithful, one hundred per cent."[49]

Explanation as Story

Fossils of extinct proboscids lie amidst the record of swamps and shallows now capped by desert sands. Millions of years after the moeritheres, the elephants had moved on, finding their niche in climates warm and cold, making their motto "Have trunk, will survive."

The structures of the elephant's big head grew big in synchrony, leading to successful foraging and fighting. In the simplest and briefest sense, elephants have trunks the better to forage and huge tusks the better to fight—and specialized throats and cheek teeth, too. Selection sculpted head morphology from what proboscid progenitors had to offer: a flexible upper lip and protruding incisors. Elephants needed a way to sweep the plain for food and load their high-held mouths. The flexible, maneuverable, powerful trunk did so—without being hindered by the presence of enormous tusks. That's a reasonable Darwinian answer to Kipling's question, an answer with no need to invoke a crocodilian deus ex machina.

How the elephant got its trunk is more than a just-so story. It's a portal to Darwin's paradigm, a reminder of the power of asking impertinent questions. A plausible Darwinist answer to "Just how did the elephant get its trunk?" considers the trunk-as-tongue, maneuvering around often gargantuan tusks—an assembly exaggerated by an arms race. By delivering big bites to an elevated mouth on an increasingly big head, it fueled huge herbivores confronted with shifting climate and changing flora. Concurrently, elephant evolution exploited the oversized features of the head in the behaviors of digging, fly-swatting, socializing, communicating, chewing, and spraying—just as Kipling pointed out. Darwinian histories, anchored in evidence and theory, make good stories. They, too, evolve as evidence accumulates, as each telling reenacts an explanation of origins.[50]

This view of life, the product of Charles Darwin's curiosity, reminds all creatures of their shared and humble origins. Darwinism underscores life's unity amidst astonishing variation, the central feature of ongoing creation. In seeking to express truth as best as evidence allows, Darwin's tales seem to lack Kipling's charm. In truth, they equally excite the imagination. For both Kipling and Darwin, story provides the stage where curiosity performs remarkable acts.

Impatience with childish thought mirrors society's impatience with scientific reasoning. But being curious—insatiably curious, adventurously curious—does not reflect immaturity. By persisting into adulthood, the inquisitiveness of the childish mind has the potential to unlock nature's puzzles. Although adults may read impertinence into a child's questioning and irreverence into Darwinian storytelling, truly figuring out how the Elephant's Child got its trunk more than entertains. It satisfies curiosity with the reward of a courageous intellectual journey successfully completed, underscoring story as scientific explanation.

Rudyard Kipling's story explained to children that when the Elephant's Child wanted grass, all he had to do was pluck it up from the ground "instead of going on his knees as he used to do." Kneeling to eat would seem to disadvantage baby elephants in a predator-rich environment. It's a good thing trunks evolved to keep their heads high and their legs straight while eating, drinking, and fly-whisking. And quite fortuitously, a trunk functioned perfectly well as a snorkel for deep wading and swimming great distances.

8

THE BEARDUCK OF BALEEN

On the Origin of New Traits from Existing Ones

> *IN the sea, once upon a time, O my Best Beloved, there was a Whale, and he ate fishes. . . . Till at last there was only one small fish left in all the sea, and he was a small 'Stute Fish, and he swam a little behind the Whale's right ear, so as to be out of harm's way. Then the Whale stood up on his tail and said, "I'm hungry." And the small 'Stute Fish said in a small 'Stute voice, "Noble and generous Cetacean, have you ever tasted Man?"*
>
> *"No," said the Whale. "What is it like?"*
>
> *"Nice," said the small 'Stute Fish. "Nice but nubbly."*
>
> *"Then fetch me some," said the Whale, and he made the sea froth up with his tail*
>
> *"One at a time is enough," said the 'Stute Fish. ". . . you will find, sitting on a raft, in the middle of the sea, with nothing on but a pair of blue canvas breeches, a pair of suspenders (you must not forget the suspenders, Best Beloved), and a jack-knife, one ship-wrecked Mariner, who, it is only fair to tell you, is a man of infinite-resource-and-sagacity."*
>
> Rudyard Kipling, "How the Whale Got His Throat"

Rudyard Kipling's Whale had an insatiable appetite and a taste for suspenders-wearing mariners adrift at sea. A particular mariner in "How the Whale Got His Throat" seemed destined for the same fate that befell Captain Ahab, who was tormented and killed by Moby Dick. This mariner, however, engineered a clever escape using the "infinite-resource-and-sagacity" at his disposal. In contrast, Ahab's monomaniacal quest sealed his doom. The common thread in the course of their misfortunes: both became very familiar with a whale's anatomy.

Melville's epic dwells on how the White Whale "by its indefiniteness . . . shadows forth the heartless voids and immensities of the universe, and thus stabs us from behind with the thought of annihilation."[1] Melville's whale is existentially overwhelming, Kipling's merely whimsically frightening—but mostly just hungry. One story is deadly serious, the other playful. But both tell of man-eating whales, the same beast that swallowed Jonah, and both share some reliable knowledge about the habits of whales.

Darwin's musings on the origins of whales find their place sandwiched between Kipling's comic relief and Melville's haunting despair. Kipling's whale talked; Darwin's whale walked. For Ahab, order in nature mocked free will; for Darwin, it endlessly created novelty. Jonah's whale merely tested his faith.

Kipling's fanciful story accounts for the novel origin of shrimp-straining, krill-eating baleen whales from an original fish-eating predator. Kipling answers the conundrum "Which came first, the toothed or the baleen whales?" The Whale nearly ate himself into starvation; a change in diet, advised the 'Stute Fish, would be his salvation. Ideally, humans would do. In anticipation of a nice, nubbly meal, the Whale stood up on and frothed the sea with its "tail," more properly named a "fluke."

Hard to Swallow

Despite a life at sea, Kipling's Whale breathed air in the manner of land mammals and it beat its fluke up and down, not back and forth as fish do with their tails. Kipling's Whale galloped on flippers and fluke, back flexing and arching, to find the Mariner, his new prey. He swallowed the Mariner, raft and all, in a big gulp—fortunately for the Mariner, without chewing.

Within the Whale's belly, the sagacious Mariner—equally astute as a 'Stute Fish—had several artifacts at his disposal and the will to use them well. Unlike Ahab sitting in his cabin aboard the *Pequod*, fatefully plotting his own destruction, the Mariner did not curse science and technology. Instead, he trusted the efficacy of his plan. With greater sagacity and far less hubris than Ahab, he plotted his escape from the belly of the Whale. He behaved most rambunctiously and even "danced hornpipes where he shouldn't," irritating the immense leviathan and causing a severe case of "hiccoughs."

Figure 8.1 Baleen "suspendered" in the mouth of a whale. Illustration by Jan Glenn.

"... and the Whale felt most unhappy indeed."

The 'Stute Fish counseled the Whale to swim to the Mariner's home shore, there to disgorge the nubbly meal.

> But while the Whale had been swimming, the Mariner, who was indeed a person of infinite-resource-and-sagacity, had taken his jack-knife and cut up the raft into a little square grating all running criss-cross, and he had tied it firm with his suspenders (now, you know why you were not to forget the suspenders!), and he dragged that grating good and tight into the Whale's throat, and there it stuck!
>
> But from that day on, the grating in his throat, which he could neither cough up nor swallow down, prevented him eating anything except very, very small fish; and that is the reason why whales nowadays never eat men or boys or little girls.[2]

The grating in the "throat" of Kipling's Whale was the structure now referred to as baleen. Real baleen, however, does not fill in the throat. A krill-eating whale opens its mouth to engorge a vast gulp of seawater laden with small crustaceous "fish" known as krill. Its baleen sits under a whale's gums like a bristly internal mustache. It strains the oceanic soup as the tongue presses water out the sides, then licks the baleen clean of the "suspendered" feast.

It's probably wise to stay clear of a feeding leviathan's mouth, lest the inrushing current prove your demise. The largest animal that has ever lived, the blue whale, traps, as Kipling describes, "very, very small fish" (mostly crustaceous ones) with its baleen. It does not feast on nubbly humans. Instead, it licks krillish nubbins from its mustache and washes them down its throat. Nubbly, but nice.

There are other kinds of whales, though, that Kipling's story ignores. Their jaws house enameled teeth, not flexible corset stays or men's suspenders. The sperm whale is one, and Moby Dick the best known among them. Melville based his fictitious story on the actual sinking of a whaling vessel from Nantucket, the *Essex*, in 1820 by a famed whale, Mocha Dick ("Mocha" was the the name of a nearby island).[3] Moby Dick stove in the *Pequod*, a full-size whaling vessel; only Ishmael was spared to tell the tale. (The captain of the *Essex* survived as well.)

Whale teeth, as it turns out, offer clues about whale origins, a subject of serious speculation yet an unsolved mystery to Charles Darwin. The Darwinian approach to solving mysteries of origins has nonetheless yielded convincing accounts of how the now legless though flippered whale with its suspendered mouth and flukish tail came to be.

 Incipient Baleen

Kipling's amusing story properly sequenced the change in diet and feeding behavior among whales: first teeth, then baleen. In Darwinian histories what works survives, and what works derives from what once worked in other ways. Suspender modification makes this point metaphorically. The structure served one function for the Mariner, another for the Whale. Suspenders, freed from the task of holding up the Mariner's trousers, and combined with strips of the raft, became a novel feeding apparatus. The Mariner looked at his raft and suspenders and imagined how to use them to his benefit. Its throat webbed, the Whale exploited its new trait to satisfy its insatiable appetite. The Whale was an opportunistic predator that switched from one high-protein, high-fat diet plan to another.

All creatures great and small relentlessly exploit opportunity. Tetrapod life reworked the skeletal elements of lobe-fins to fashion arms and wrists. With articulated limbs, first lobe-fins, then tetrapods, exploited diverse habitats from the water's edge onto land.

The lobe-fin was an "incipient" limb—again, not actually an arm or a leg but the antecedent structure. When opportunity knocks, life evolves. The Mariner's suspenders and sliced rubber raft were incipient baleen—present and ready for transformation. Equipped with strainer gums, whales could reap the harvest of the sea. First, of course, they had to discard the accoutrements of land-living: legs. All that evolutionary work fashioning limbs from fins only to be undone!

In the struggle to survive in terrestrial habitats, forms changed. Swimmers became walkers. Gaits shifted from the sinuous sideways slither of amphibious salamanders (the left-right wiggle inherited from fish), legs sprawled to the side. Speedy reptiles tucked their legs beneath their torsos, and mammals mastered dorso-ventral flexing—the up-and-down motion exaggerated by a bucking bronco.

Creatures in the air-breathing, land-loving lineage leading to whales lost their hard-won hind limbs while their forelimbs transmogrified to flippers. Runners became swimmers. Ancestral whales' backbones arched powerfully to flex the fluke, a motion improved by the disconnection of the hind limbs from the backbone.

Gulping while aqua-galloping worked well as predaceous mouths began straining to feed. Fluke propulsion achieved remarkable speeds in the chase for tiny krill. In the sea, straining by baleen enormously expanded the availability of food for energetic predators of great size. The Mariner's suspenders unlocked a treasure trove of food of such abundance it could satiate the hunger of the world's most enormous beast. Kipling's Whale lost a meal but gained a way to bulk up.

The Bearduck

Kipling wondered, "How did the whale get its baleen?" In Darwinian guise the question becomes "Of what use to an ancient whale might baleen have been before it had the size and structure modern leviathans depend upon to strain the ocean's plankton for food?" In other words, Darwin wondered what might have been the incipient trait.

Variation and scaling figured in Darwin's answer. Long before such evidence accumulated, Darwin speculated on how to make a whale from something altogether different. He did so by invoking the idea of "incipient organs," explaining the potential utility of a development at its earliest stages to become a novel adaptation in "finely graduated steps, each of service to its possessor." He continues:

> The Greenland whale is one of the most wonderful animals in the world, and the baleen, or whale-bone, one of its greatest peculiarities. The baleen consists of a row, on each side, of the upper jaw, of about 300 plates or laminae, which stand close together. . . . The extremities and inner margins of all the plates are frayed into stiff bristles, which clothe the whole gigantic palate and serve to strain or sift the water, and thus to secure the minute prey on which these great animals subsist. The middle and longest lamina in the Greenland whale is ten, twelve, or even fifteen feet in length.[4]

The Greenland whale feeds by licking minute prey strained from the sea by the bristles in its mouth. It swims to eat and eats to swim; its whale of a tongue is its principal eating utensil. The problem for Darwin was to surmise how such "organs" (the baleen bristle structure or "whale bone" and the giant tongue) might have served a useful function when still in an incipient state. He reasoned that the incipient organs first existed at small scale. Enormity came later.

Darwin's contemporaries found fault in the notion of incipient organs. Modern-day students of evolution experience a similar problem. The logic is concise: What good is a _____ that is not yet a _____? What good is a baleen-draped mouth if the baleen is small and inconspicuous?

In order to imagine a plausible answer, Darwin looked for feeding behavior analogous to that of the whale: swimming and filtering at the same time. Ducks do it. He responded to one of his most troublesome critics, St. George Mivart, by proposing the image of a proto-whale sifting the water like a duck. The beaks of dabbling ducks are edged with a laminar structure that resembles a fine-toothed comb, the pecten. Pond water comes in, the mouth closes, pond water goes out, and the fine-toothed comb strains out the duck food. Duck enlargement would entail embiggening the pecten—or "incipient baleen." In proposing a beginning for baleen, Darwin wrote:

> With respect to the baleen, Mr. Mivart remarks that if it "had once attained such a size and development as to be at all useful, then its preservation and augmentation within serviceable limits would be promoted by natural selection alone. But how to obtain the beginning of such useful development?" In answer, it may be asked, why should not early progenitors of the whales with baleen have possessed mouths constructed something like the laminated beak of a duck? Ducks, like whales, subsist by sifting the mud and water; and the family has sometimes been called *Criblatores*, or sifters. I hope that I may not be misconstrued into saying that the progenitors of whales did actually possess mouths laminated like the beak of a duck. I wish only to show that this is not incredible, and that the immense plates of the Baleen in the Greenland whale might have been developed from such laminae by finely graduated steps, each of service to its possessor.[5]

Darwin never identified such a creature, living or fossil. He only wished to demonstrate the plausibility of "finely graduated steps" from an antecedent structure, each step being of good "service to its possessor." Fine laminae did exist in nature, and one might suppose that they existed in the mouths of ancestral whales. The immensity of the whale implied scaling fine laminae to enormous size.

As mentioned, duck bills have elastic, comb-like, toothish structures, and Darwin had carefully observed them. Scale a duck head to whale size and voilà, baleen happens. Or at least a really scary-looking duck:

> The beak of a shoveller-duck is a more beautiful and complex structure than the mouth of the whale. The upper mandible is furnished on each side (in the specimen examined by me) with a row or comb of 188 thin, elastic lamellae, obliquely bevelled so as to be pointed, and placed transversely to the longer axis of the mouth. They arise from the palate, and are attached by flexible membrane to the sides of the mandible. . . . In these several respects they resemble the plates of baleen found in the mouth of the whale. . . . If we were to make the head of the shoveller as long as that of the Balaenoptera, the lamellae would be six inches in length,—that is, two-thirds of the length of the baleen in this species of whale."[6]

To Darwin, scale was the key to understanding the possible origin of baleen from a micro-scale, incipient organ already being used in feeding while swimming. Serendipitously, Darwin knew of a bulky creature that skimmed as it swam: the bear. Bears sometimes eat vast quantities of insects and are perfectly capable of chasing after them in the water. Thus, Darwin proposed the example of a swimming bear feeding like a shoveller duck:

> In North America the black bear was seen by Hearne swimming for hours with widely open mouth, thus catching, like a whale, insects in the water. Even in so extreme a case as this, if the supply of insects were constant, and if better adapted competitors did not already exist in the country, I can see no difficulty in a race of bears being rendered, by natural selection, more and more aquatic in their structure and habits, with larger and larger mouths, till a creature was produced as monstrous as a whale.[7]

A contemporary critic, John Morris, in an otherwise favorable review of Darwin's *Origin of Species*, demanded evidence that the gut of the black bear swimming, mouth agape, was in fact full of insects. Morris reasoned that a bear could not feed like a whale because a bear lacked baleen. He asked rhetorically, "How could it possibly have fed on such sea insects, when its mouth is not furnished with the beautiful apparatus which enables the whale to retain its tiny food while it ejects the water that contained it?"[8]

Morris could not fathom a series of finely graduated steps from bear tooth to baleen. He accused Darwin of using a suspect fact simply because it suited his theory.

No one today proposes the descent of whales from the same ancient creatures that likely gave rise to bears, dogs, cats, sea lions, and weasels—the modern carnivores. Darwin's imagined consequences of bears swimming with open mouths to snag insects strained credulity, so much so that he greatly abridged this passage in later versions of his *Origin* where only the first sentence remained. He never gave up, however, the idea of incipient traits and finely graduated steps, each of service to its possessor. In private correspondence he held on to his affection for the notion of natural selection being quite capable of turning bears into something

Figure 8.2 A leisurely paddling bearduck. Illustration by Jan Glenn.

 as monstrous as a whale. He simply imagined the consequences, generation after generation, of bigger and bigger-mouthed bears being better and better able to skim insects from the water's surface like a duck.[9] Let's credit him for never fully abandoning the wonderfully imaginative bearduck of baleen as a hypothetical example.

Tied to an understanding of incipient traits is the notion of transitional (or intermediate) forms. There are pitfalls to avoid when thinking in oversimplified ways about transitional forms because they can only be recognized in retrospect. Similarly, identifying an incipient trait or organ depends on placing it in relationship to a subsequent adaptation. Nevertheless, Darwin provided a plausible scenario and identified reasonable analogues in order to account for how the whale got its throat without any assistance from a sagacious Mariner (or divine designer). Moreover, his reasoning pointed out the utility of a new structure (baleen) even at the beginnings of its development. His approach has proven most useful to de-puzzling origins.

Legless Bearducks

In Kipling's story the sagacious Mariner intelligently designed the modifications that gave the Whale a new adaptation. Animal evolution, of course, works quite differently. A good Darwinian story does have at least one element in common with the Mariner's approach: old parts put to new uses. New combinations often in changed proportions become novel adaptations. That is the point of Darwin's duck-and-bear analogy to whales.

The anatomical parts exploited by natural selection to create novelty are present in the precursor creature: lobe-fin bones to limb bones and gill arches to jaws, for example. Tracing the history of these changes constitutes an explanation of a creature's origins—Kipling's whimsical goal as well as Darwin's serious one.

Those who have followed Darwin's approach have found similar forms among creatures of the past and whales of the present at the scales of both molecules and body parts. Their work has yielded a convincing story of whale evolution and a cornucopia of fascinating, even entertaining images anchored to fossil evidence. What makes a whale a whale? How did it become so? Bearducks never existed, but walking whales did.

Paleontologists claim to have found evidence of whale legs with specialized joint structures at the ankle. These anatomical structures link whales to ancient even-toed ungulates whose descendants sport not flippers but hooves.[10]

The bones of fossil whale limbs resemble those of animals that evolved a tiptoe stance, the posture of a ballerina on pointe, aptly animated as dancing, four-toed hippos in Disney's *Fantasia*. When the groups had yet to branch apart, they probably walked (and danced) a bit more flat-footed.

As with fish to land, there is a puzzle to solve in the transition from land to sea among the whale's ancestors. Traits that change very little often link the intermediates: for example, enameled whale teeth, the raw material of scrimshaw. Even the baleen whales grow hard teeth during embryonic development. These temporarily grown teeth are then reabsorbed. Baleen-feeders are tooth-absorbers. If the tooth of a creature has the structure of a whale tooth, then the creature is likely a whale or a close whale relative—even if the animal had legs.

Darwin derived a hypothesis regarding the origin of baleen among whales by an analogy to bears feeding like ducks. His idea held little credibility in his own time and as a specific evolutionary history fares no better today. Nevertheless, whether labeled as pre-adaptations, precursors, or incipient organs, the variations among existing traits, winnowed by the struggle to survive, yield new and novel forms. Flexible behavior and variable diet are often key to the exploitation of new environs where novel variations might flourish.

The Mivart Objection

Darwin wondered not only about baleen but no doubt also about flukes and flippers, as in: "How did the whale lose its legs?" The notion that essential adaptations had precursors that functioned in some other way raised a vexing question: "How can a partially formed trait or 'organ' be of beneficial use at each gradation of its development?" Assuming legged whales slithered from shore to sea and that today's whales gallop sinuously and gracefully through the ocean, what did the intermediates do? Perhaps they could scramble and paddle in otter fashion.

Those attacking Darwin believed "incipient organs" to be absurdities. They believed that unbridgeable divides separated distinct anatomical designs such as hoofed and fluked and argued in favor of independent origins in dispersed geographic centers of creation. The Darwinian claim of common descent looked loony to them. The Creator created creatures harmoniously adapted to their environs in separate places and that was that. The absence and impossibility of intermediates made their case.

St. George Mivart articulated the incipient organ objections so forcefully that even some of Darwin's allies began to entertain doubt about the effectiveness of natural selection. Darwin's opponents were neither unintelligent nor illogical. At the least, concluded Mivart, natural selection would work to expunge an incipient organ the same as it would a defective one. In both cases, the disadvantage of poor performance would be apparent. How could one and the same process both expunge and preserve, both cull and select, variants of a trait? Would not the weak performance of early stages lead to extinction rather than to improvement across generations?

The classic challenge to Darwinism in this regard was the structure of the eye. How could an organ so exquisitely fashioned in every detail spring into being piecemeal? The allusion to the Bible in Mivart's title, *On the Genesis of Species*, was not accidental. Neither was the parallelism in rhetorical form to Darwin's own title, *On the Origin of Species*.

Mivart's criticism of incipient organs (gum bristles before baleen) or its extrapolation, incipient species (lobe-fins before tetrapods), bedeviled evolutionists. Darwin responded squarely, sometimes strengthening his arguments with new examples of variation, sometimes yielding ground unwisely in order to accommodate false (as it later turned out) claims about the age of the earth. Darwin's idea implied time's unfathomable vastness. A contemporary physicist's calculation disallowed ample time for natural selection to work its magic. Or so it seemed at the time.

Lord Kelvin convinced the nineteenth-century world that the earth had cooled from a molten state in just several million years. This truncated timeline encouraged Darwin to elaborate and endorse ideas about how the "use and disuse of parts" (behavior) and the "conditions of life" (environment) might accelerate evolution. He came to sympathize with the idea that the severe demands of survival in one life might influence the vitality of the next generation. Physics, by estimating the time needed to cool the earth from an initial molten state, had robbed him of the immense spans of time his extrapolations of origins from incipient species depended on.

Many decades later, radiometric dating proved the "cooling clock" to be wildly inaccurate. Darwin never knew that, in truth, time's vastness squared nicely with his bedrock proposal: descent with modification by means of natural selection, without recourse to accelerating mechanisms.

How *does* one explain the early beginnings of a useful development? If the existing trait is of great adaptive value, then how can this use be abandoned? The Mivart objection applies to both situations: structures half-formed or half-abandoned. In either case, modification appeared to be a liability. Imagine the predicament of being bombarded with such sensible questions by an educated, skeptical audience. Darwin did not claim that a stranded mariner had sliced up a raft and a pair of suspenders, fitted this mesh to the mouth of a whale, and thus invented baleen. Nevertheless, Many of his critics found his thinking scarcely more credible. Mivart's objection to the idea of incipient organs (the uselessness of a mostly blind eye or "half a wing," for example) persuaded even Alfred Russel Wallace, coauthor with Darwin of the first scientific paper on natural selection and popularizer of the phrase "survival of the fittest," to admit, "How incipient organs can be useful is a real difficulty."[11]

The Piecemeal Eye and the Stickleback Spine

Of what good, argued Mivart, were incipient eyes, eyes not yet eyes? Mivart's objection to the viability of an intermediate eye is important to refute because the same logic applies to every other organ and behavior from elephant trunks to whale flukes, from elbows to ankles, from inner ears to kidneys. Do eyes exist along a continuum resembling the evolutionary journey from incipient trait to fully functioning organ of vision? The arthropod eye, the vertebrate eye, and the cephalopod eye are all very similar in function but different in structure. Each sees a part of the world in its own adaptive way, responding to forms perceived in murky depths or brilliant sunshine.

How could an organ so complex as the human eye spring into being piecemeal? Today, the "piecemeal eye" has been documented as effective at every degree of sophistication. Each step of increasing complexity serves its possessor well. Among animals, examples are known of variations in photosensitive reception as well as lens structure. Earthworms, for example, can detect light; they prefer darkness. But they do not form images of their surroundings; they cannot see fishhooks. Selection's tinkering amplifies such primitive features and converts them to new purposes. It does so by favoring spontaneous mutations, especially in the genes that regulate the expression of other genes.

For example, light-sensing, light-gathering, and light-focusing structures are readily observed across a host of living creatures. Among all of them, the same regulatory gene, *Pax6*, is at work. It controls aspects of nervous system development including the eye.[12] The *Pax6* gene has been part of the tree of life for hundreds of millions of years, from before the vertebrate and invertebrate lineages had branched apart. *Pax6*—a genie in the genome—rebuts Mivart. The gene is a molecular fingerprint, forensic evidence of common descent from a very distant ancestor. Just about every way an organism can detect light serves well in one context or another, from light-sensitive eyespot to fully imaging organ. Variants of *Pax6* play a role in them all.

The transition from incipient eye to clear vision started with pigmented cells, proceeded through cup-like structures, and led to enclosures capped by lens structures of varying contrivance. Survival welcomed the chance to detect light to whatever degree and by whatever organ selection could fashion. The human eye can detect depth, color, shape, and motion with remarkable acuity. Familiarity with human eyes blinded Mivart's understanding of how eyes very unlike human eyes worked perfectly well for organisms very unlike humans.

The expression of eyes across lineages and through time mirrors the gestation of limbs. Turning on or turning off *Pitx1*, another gene from the dawn of Animalia, plays a role in limb development in the bodies of many species. For example, turning off *Pitx1* in the hip region of the three-spine stickleback fish suppresses the growth of pelvic spines. That's a good thing in muddy-bottomed freshwater habitats where dragonfly larvae prey upon wandering sticklebacks. Smooth backs give predatory dragonfly larvae a little less to grab on to. In open water, stickleback pelvic spines make for a good defensive strategy since predator fish avoid hooking their mouths on stickleback barbs.[13] For an open-water stickleback, *Pitx1* is a turn-on.

Pitx1 suppression may work in a similar fashion to block hind limb development in whales. Of course no dragonfly larvae threaten whales, so some other selection pressure gave limblessness its advantage in this case. Swimming after krill with mouth agape, the body streamlined for speed, fluke cranking away, comes to mind. A whale's body is always narrowest just before the fluke. Legs would only get in the way.

Darwin's Dilemma Reconciled

The variability and evolution of both eyes and limbs demonstrates the viability of intermediate—hence incipient—organs (as well as the powerful role of regulatory genes subject to mutation). "Primitive" stages have vital uses at the time of their flourishing; incompleteness is in the eye of the beholder.

Darwin outlined plausible reconciliations to the conundrum of incipient organs. He understood how descent, amidst an intense struggle to survive, proceeded step by virtually imperceptible step from blindness to eagle eyes, from running to flight, from ambling shore dweller to deep sea diver, from a mouthful of who-knows-what to fully formed baleen, of service not only to leviathans but also to Victorian era women's undergarments. Darwin doubters conjure "gaps"—the unsuccessful search for fully functional intermediate forms. "Gapologists," in keeping with the Mivart objection, conclude that parts absent from the whole cannot work, whether the parts are of a limb or of a molecule. Gapologists may confuse the public, but they fare no better than St. George Mivart in solving puzzles of origins or refuting Darwinian solutions.[14]

Symbolized in the good of an eye-not-yet-an-eye, the limb-no-longer-a-fin, Darwin's dilemma stands reconciled. The presence of ancient genes in dramatically different organisms—and variation in their regulation—validates the claim of descent from a common ancestor to a degree Darwin could not imagine. As he hypothesized, there is potential utility for a complex organ at intermediate stages of development. In "finely graduated steps, each of service to its possessor," new traits, organs, and species evolve.

Evidence obtained by modern genetics and from the fossil record combine to tell the story of incipient baleen. The careful analysis of enamel-making sequences of DNA in the genomes of extant whales has exposed the machinery for making both baleen and teeth. This finding reinforces the conclusion that there once existed an intermediate whale sporting both.[15] Specimens discovered in Oregonian rocks from the ocean floor of 27 or so million years ago confirm this expectation. Paleontologists claims that the morphological features observed in the palate of the fossilized skull of *Aetiocetus weltoni* resemble those for nourishing baleen found along the sides of the palates in living whales. The skulls of aetiocetids appear to offer a perfect example of incipience unknown to Darwin and negate any need to hypothesize bearducks.[16]

Extraordinarily strange plates of baleen hang from the upper jaws of the most giant creatures ever to inhabit the globe. They evoke the image of the suspenders and strips of raft crafted by Kipling's sailor. Baleen is made of keratin, a ubiquitous skin protein, and grows throughout a whale's life. Baleen is split fibers of keratin on a gigantic scale. Keratin adds toughness. It's the material of horns, nails, and claws. It's the hardener of bird beaks and tortoise shells. Hair, feathers, and scales are keratin-based. According to its dictionary definition, "keratin is a fibrous protein that is the main constituent of the outermost layer of the skin, nails, and hair. It is resistant to damage from chemical and physical agents."[17] Keratin is incipient baleen.

Baleen grows and baleen wears out. It needs no trimming but might be thought of as a bushy under-the-upper-lip mustache, scaled to size along with the rest of the whale, no more out of place than the straining strands of a duck's bill, as Darwin noted, and not much more problematical in origin than a rhino's keratin-rich horn.

How the camel got its hump, the whale its throat, the elephant its trunk, the giraffe its neck, the leopard its spots, and endless more problems of the same kind demand imaginative responses. Change the regulated growth of a bit of bristle beneath the gum, and the same substance that makes mustaches lurches in the direction of baleen. The bearduck, an incipient baleen whale of the imagination, reminds us not to discount what evolution can accomplish. Darwin's basic claim, the origin of species from descent with modification by means of natural selection, remains sound, even if bearducks in the proportions of whales never happened.

9

THE SAGA OF MOOSHMAEL

The Logic of Family Relationships

One day
Morris the Moose
saw a cow.
"You are
a funny-looking moose,"
he said.
"I am a COW.
I am not a MOOSE!"
said the cow.
"You have four legs
and a tail
and things on your head,"
said Morris.
"You are a moose."

Bernard Wiseman, *Morris the Moose*

Meeting a cow, Morris the Moose began his thinly veiled and clumsy attempt at flirtation with a backhanded compliment followed by a logical fallacy.[1] Cows do not appreciate illogical wooing. Furthermore, a cow does not like to be told that she is a funny-looking moose. A genteel buffalo, perhaps, but not a moose.

Morris reasoned: If a creature is a moose, then it has four legs, a tail, and things on its head. The creature he encountered had all of these features and was therefore a moose. Morris must be forgiven his fallacy. He affirmed the consequent. True, moose have these traits. True, the creature he met one day had these traits. But as the reader recognizes, the cow is a cow and not a moose. Many animals have things on their head, four legs, and a tail; not all of them are moose. A cow, however, *is* exactly the same as a moose except for the differences. Indeed, a cow and a moose are completely different except for their similarities. Cow and moose similarities and differences, in terms of shared traits, are useful elements of playful logic. They are also the raw data for drawing evolutionary inferences.

Befuddled by Morris's seemingly cogent argument and deceptively impeccable logic, the "other moose" attempted to negate his conclusion with a new fact: "But I say MOO!" Quickly Morris mooed too, and the argument continued.

Morris and the cow next debated the relevancy of giving milk. Morris felt that milk-giving was secondary to other, more important features that determined moose status. His newly encountered four-legged, tailed, and things-on-head friend was a "moose who gives milk to people."

Morris failed to sympathize with any of the cow's points. In exasperation the cow exclaimed, "But my mother is a COW!" and therefore, as her mother's child, she must also be a cow. Cows birth cows.

Morris was too sharp to fall for such illogical heresy. "You are a MOOSE," he repeated, and then reasoned quite clearly, "So your mother must be a moose too!" If Morris were to extrapolate this line of reasoning across many generations, then the first organisms to crawl onto the land and lay eggs would also have been moose, but without things-on-the-head and other modern moose accoutrements.

Morris and the cow went in search of an independent adjudicator of their dispute. They found a deer and quickly concluded it was another moose and therefore potentially biased. When they met a horse, the horse exclaimed, "Hello, you horses, . . . what are those funny things on your head?" No horse has horns or antlers, of course. Deer are very similar to moose (actually from the same family, Cervidae), cows a bit less so, and horses perhaps something else altogether. When they put their heads together and reflected on their images in a nearby stream, the four quadrupeds recognized and accepted their differences. They were wise quadrupeds indeed, very evolved. Morris was moose enough to admit that he had "made a MOOSEtake!"

Morris the Socratic Moose

Gareth Matthews in his book *Philosophy and the Young Child* reminds us that ten-year-olds have the potential to mimic Socrates by asking disarmingly simple questions that call for the reexamination of things thoughtlessly taken for granted.[2] Children delight in literature that plays with the kinds of questions they find both silly and interesting. "What is a moose?" (or a fish or a whale or an elephant) is just such a question. As pointed out by Matthews, Morris the Moose faced the problem of "mooseness." Having related the story as Bernard Wiseman tells it, I now retell it here, with apologies to the original author, and in keeping with Matthews's analysis of the significance of classification criteria to thinking and reasoning about the essential as opposed to incidental characteristics of "mooseness."

Morris and the cow found a deer. Seeing at once that both his moose and cow friends had erred, the deer proposed a higher taxon within which to classify both moose and cows: "deer." The category "deer" referred to all those creatures with four legs and things on their heads. Cows and moose are different types of specialized deer. Had the deer received zoological training, he would have viewed his hoofed friends as members of the order Artiodactyla, the even-toed ungulates—cows, deer, pigs, antelope, giraffes, peccaries, hippos, and camels. (This order figures prominently in the tale of whale origins.)

The threesome, Morris, the deer, and the cow, approached a horse, who greeted them with a hearty "Hello, you horses!" Apparently the horse had not the benefit of a higher education from which he might have learned to distinguish properly between even-toed ungulates and odd-toed ones such as himself. As all four bent over to sip water from a stream, Morris noticed in their reflections how unique he looked in comparison to the others and revised his conclusion accordingly. Cow was cow, moose was moose, and that was the end of the story.

Well, not quite. Some time ago the news media picked up the story of another moose, nicknamed Bullwinkle, who wandered out of the backcountry onto a Vermont farm seeking a mate.[3] Perhaps having recently read *Morris the Moose*, it pined after a cow named Jessica. The story became popular, and Bullwinkle and Jessica's romance was turned into a children's book, indicating that confusion as to proper species classification troubles not only Morris but also evolutionists and real-world moose (at least Vermont moose).[4]

The moose-cow dilemma exists in many guises. As Gareth Matthews playfully asks, "Is a bicycle a tricycle without one of the wheels? Is a snake a lizard with no legs? . . . Whimsical questions of this sort can introduce a thoughtful discussion of the practical and philosophical problems of taxonomy."[5]

Such questions more than amuse. They parallel those posed by paleontologists: "Was *Archaeopteryx*—the feathery winged creature of tens of millions of years ago—a reptile with feathers or a flightless bird with reptilian claws and teeth?" Decisions on how to classify moose, cows, fossils, birds, dinosaurs, reptiles, toys, and household objects depend ultimately upon how humans intend to use their knowledge.

What a conversation Morris the Moose might have with a paleontologist! Which fossils with things on their head—the Irish elk, for example—are moose and which are non-moose? Think about the possibility of some Doris the Whale, in mimicry of Morris's logic, trying to explain to her ancient land-dwelling ancestors that they must be whales because their progeny became whales. Beach-slithering, tailed and limbed, long-jawed animals with no spermaceti in their head might beg to disagree. Perhaps they would even be offended, preferring to think of themselves as primitive hippopotamuses.

As it turns out, there is a veritable paleontological cottage industry of studies about how the whale lost its limbs and got its fluke. In several of these studies fossil finders write about the walking whales of Pakistan and India and even Egypt. Imagine these beasts walking among snorkeling, swimming proto-elephants in what would become Egypt or basking on shorelines destined to rise as Himalayan peaks.[6]

Picture a modern whaling Morris, the moose equivalent to Melville's Ishmael. "Call me Mooshmael," he writes at the beginning of his story. Mooshmael knows well the literature on the anatomy of the whale, both fossil and living, and understands cetology from a Darwinian point of view. He's better educated than Melville's Ishmael. Nevertheless, he does not accept that a whale is anything other than a funny-looking moose.

One day Mooshmael set sail on the ship *Artiodactyl* and shortly thereafter spied a "whale" off the port bow.

"Ahoy there!" shouted Mooshmael. "What a strange-looking moose you are!"

"I'm Doris and I'm no moose; I'm a whale," the toothed cetacean shouted back.

"You must be mistaken. I see that you are a legless moose that swims, yet breathes air and gives milk, just like a moose," Mooshmael replied.

"I am a whale because my nostrils are on top of my head, where your antlers stick out," Doris explained.

"No, you are a moose because you use your head to batter things, the same as I do during the season of the rut. And your blood is warm, like mine," said Mooshmael.

"That may be," admitted Doris, "But I can never walk on the land like a moose, and I swim with a flattened tail that pivots up and down on my lower back."

"So you say now," Mooshmael retorted. "But I happen to know about your ancestors, and they most certainly had legs and walked. In fact, they most certainly had feet and ankles like mine—like a moose. *Since they were moose, you must be a moose.* I strongly suspect you are my long-lost cousin!"

"You are the one who is mistaken. *They must have been whales because I am a whale*," reasoned Doris.

Mooshmael invited Doris to swim alongside the *Artiodactyl*. He opened his ancient fossil-photo album. "I like to keep pictures of all the branches of my family tree for many generations past," Mooshmael told Doris. "We moose, and all the members of our extended family, are quite proud of our running ability. It comes from our very special and unique ankle bones. And, of course, we moose also swim quite well, as I see you do."

"Ah, running," sighed Doris. "That's what makes you a moose and me not a moose."

"Not so fast," said Mooshmael. "I admit you don't look like a marathon moose to me. Are you sure you lack a world-class double-pulley ankle joint and the even-toed hoof that's been in the family for so many generations? My hippo brethren, obviously not built for sprinting, still grow four toes, but as you can tell, I've grown accustomed to having just two on each foot."

Mooshmael returned to looking at his family album. "I see here a branch of the family with a horizontally flattened tail and four-webbed limbs. It's Very Great Uncle Basil (a *Basilosaurus*), a moose you resemble quite closely, if I do say so."

Doris looked at the picture of Very Great Uncle Basil. Mooshmael was correct. It was indeed huge with a streamlined body, flipperish forelimbs, and tiny hind limbs. Definitely not a walker, let alone a runner—some sort of weird mutant whale. "You said that running makes the moose a moose. Very Great Uncle Basil looks to me like a whale with feet too tiny for running."

"Good point," admitted Mooshmael. "But I think the running is in the bones, no matter how much they have faded away. Look here at Very Great Aunt Rodhie (a *Rodhocetus*). She lived some time before Very Great Uncle Basil. Take a look at her hind legs!"

Doris stared at the picture of Very Great Aunt Rodhie's hind leg. To her astonishment, the ankle bone looked exactly like the one Mooshmael had bragged about: two rounded ends, just right for a runner's leg motion. Yet, according to a note below her picture, Very Great Aunt Rodhie was remembered as a champion swimmer. She had a long body and a powerful tail, though not flattened like a fluke. Her leg and arm bones were limbs, not flippers, but they were flattened and paddle-like. Doris thought of her friends, sea otter and sea lion. Very Great Aunt Rodhie reminded her of them. "I'm getting confused," muttered Doris. "Are you trying to convince me that I am a moose or that you are some kind of whale?"

"I guess it makes little difference," Mooshmael reflected. "You could think about our relationship either way. I prefer, of course, the why-can't-we-all-just-be-moose point of view. Let me show you Very Distant Cousin Amble's (an *Ambulocetus*) best picture. No one knows whether Very Distant Cousin Amble preferred to walk or swim, but you can tell that she had a hoof-like toe."

Doris studied the picture carefully and saw clearly the flexible, strong wrist and ankle joints. The mouth and long snout were those of a fish-eater. Its limbs looked like they could take strong swimming strokes, but the picture showed Very Distant Cousin Amble ambling along near the shore.

"Okay, Mooshmael. I think I know now where you are mistaken. It's those ears. They are similar to whale ears, something you moose would never understand. Whale ears work in water—we've got thickened bones to carry sound and spaces in our heads that keep what the left ear hears separate from the right. It's how we find our way by listening in the dark and murky ocean depths. Very

Figure 9.1 Mooshmael sharing family photos with Doris. Illustration by Jan Glenn.

Distant Cousin Amble must be a whale because she listened like a whale and lived like a whale—at least sometimes."

"Ah, so you admit: whales walk! If they walked, maybe they ran, too—just like a moose! Here, look now at this picture of Very Very Great Uncle Pakky (*a Pakicetus*) from Pakistan. You can see his ear very clearly—and look at those teeth, unquestionably the right tools for chomping on fish," said Mooshmael.

"See, you are plainly being misled," said the whale. "That ear of Very Very Great Uncle Pakky would never work for a whale. Why, it even lacks sponge-filled sinuses to isolate the sounds heard by the right and left ears! Those are the ears of a landlubber, whatever the rest of the skull and the teeth may have in common with whales—and I don't see much."

"Look more closely," Mooshmael said, "and you will observe that there is a thickened bone able to transmit sound underwater."

"Or maybe Very Very Great Uncle Pakky simply enjoyed resting his head on the ground and listening for vibrations through the earth for prey scampering about," said Doris.

"Well, then, pay attention to the eyes, snout, hips, shoulders, joints, and limbs. What a runner! Your ancestor and mine, Cousin Whale, and obviously an even closer relative of Very Distant Cousin Amble," concluded Mooshmael.

Doris had been thinking. "Not so fast, Mooshmael. Very Very Great Uncle Pakky has the eyes and ears of a predator. Are you suggesting you moose, who now feast on lilies and pond plants, and all your extended family that graze and browse, are the Very Great Grandchildren of a meat-eating predator? Has the hunter become the hunted? You have tried to convince me that we whales have shifted our shape in its 'external respects' and lost our limbs. If I am to believe you, I must also believe that you moose have switched teeth and changed your diet completely. I find your claim to have so entirely altered your 'internal respects' astonishing. You might as well tell me that the hippo is my closest living relative," snapped Doris.

"It might well be," said Mooshmael, as he ruminated some more on Doris's argument.

The teeth—those mutable teeth, knife-sharp to tear apart fish; cross-ridged and broad to grind down plants. Teeth trapped Mooshmael within his net of inferences; teeth cast doubt on the authenticity of his family album. Teeth could not be trusted.

"Teeth come and go. Only the bones remember," Mooshmael asserted philosophically.

"Whether or not dentition varies with descent, I do know that my closest living relative is not the hippopotamus but my lovely and toothless cousin Helene, the baleen whale. Her smile could launch a thousand ships. She is truly like family to me," Doris said lovingly. "I think I hear her whale song now, calling in the distance."

Mooshmael was stumped. Baleen indeed; nothing he could recall about the odd traits in his extended family appeared to resemble baleen. Once before Mooshmael had noticed something weird about his toothless cousin Helene. He had always wished to ask about what appeared to be a gigantic mustache under her upper lip but knew it would be very rude to do so.

Pig, cow, sheep, goat, camel, antelope, and deer had all shared a journey at one time or another with Mooshmael aboard the *Artiodactyl*, and he knew that they all browsed or grazed with proper teeth. No one, thank goodness, would ever think of making women's undergarments from parts of a moose's mouth. As he now realized, there were whales with teeth and whales with baleen—corset frames and bustle stays—in their mouths. He thought for a moment and then remembered reading about this question. It was something about women's undergarments that triggered his memory of men's suspenders stuck in a whale's throat . . . or was that a woman's corset hanging there? As Mooshmael once again paused to ruminate, Doris the whale swam away, her own brow furrowed in thought, "one broad firmament of a forehead, pleated with riddles."[7]

Mooshmael awoke the next morning to a surprise in his mailbox. An elderly great-great auntie had passed away, leaving much of her inheritance to Mooshmael. Auntie Indo (short for *Indohyus*) hailed from the Raoellidae family of India. Mooshmael had her legs and ankles, but her smile flashed teeth different from his. A faded tintype photo came with the letter from her attorney. Auntie Indo had been born nearly 50 million years ago. She had lived a long life, and her great-great-great and even greater grandchildren, grandnieces, and grandnephews lived all over the world. His ancestors' success made Mooshmael feel very special.

The letter indicated that Mooshmael was to find Auntie Indo's long-estranged distant niece Doris, and Helene, her second cousin once removed (or something like that), with whom Mooshmael was to share the inheritance. Imagine Mooshmael's surprise: the strange-looking moose he had just met might be a distant relative. And now, thanks to a surprising shared inheritance, they were both rich!

The package also included her dental records and the preliminary results of an autopsy. Auntie Indo, it was said, had some uncanny hearing abilities. The autopsy showed skull features that might explain why, as well as some interesting bone density results. Indeed, Doris had mentioned hearing Helene's distant voice when Mooshmael could not. He then looked over the dental records. Where had he seen teeth like these before? *Aha!* he thought. *When Doris opened her great mouth to speak! Both Auntie Indo and Doris boasted copious cusps of scrimshaw-ready enamel!* Doris's teeth did have the more elegant and modern peg-like look. Auntie Indo's looked more cusped and primitive, as would befit someone of her age.

Mooshmael remembered stories of how Auntie Indo would spend hour after hour wading and snacking. According to the dental records, the composition of her teeth indicated a lifelong diet of foods from aquatic habitats. How he too loved to browse while knee-deep in cool waters. Quickly he realized that he must find Doris and Helene and tell them the news about their shared inheritance from Auntie Indo.[8] He tacked the *Artiodactyl* and sailed in the direction of his newly found "cousin."

What Morris Got Right

The imagery of terrestrial hunters becoming oceanic predators—the whale's evolutionary story—reveals how flukes happened and hind legs vanished. Hearing adapted to waterborne sound; noses and eyes migrated across the head. Jawbones and skulls filled with oil. And all through the ages the "intermediate" forms put their "transitional" traits to perfectly good use. Mooshmael's ancestors were a diverse and successful lot.

In the whale's story, eyes peering over the surface of the water migrate from a crocodilian snout to the side of the head, each eye isolated from the other. The nose no longer flares at the tip but transmogrifies to become the blowhole on the top of the head. The terrestrial fish-eater's cusped and grinding molars give way to simple prongs of teeth, ideally suited for the Nantucket whaler's scrimshaw art, ill-suited for any chewing; the ultimate big gulp. Teeth disappear altogether among adult baleen whales. (Although enameled teeth do appear during embryonic development, they are reabsorbed and disappear.)

From a Darwinian perspective, moose and cow ambulate in harmony. Both descend from even-toed ungulates, the cause of Morris's confusion. The horse made a good point. Horses and rhinos made their odd-toed journey into hoofdom in their own way, long before either cows or whales came to be.

"What is a whale?" Crossing a threshold makes a whale a whale, a moose a moose, or even artiodactyls (the even-toed ungulates) artiodactyls.[9] The threshold is not easy to define. Sometimes a small number of bones will do the trick. Observing mate selection works well, but not with fossils. "I know a cow when I see one. And a moose, too." It's what a cow's got that a moose has not that makes the difference. Its endearing call, "MOO," for instance. Then again, "MOO" may just be an endearing way of calling "MOOOOSE!" Jessica might think so.

10

THE HIGGLEDY-PIGGLEDY WHALE

Leviathan's Walk, Paddle, and Gallop to the Sea

You must not, in every case at least, take the higgledy-piggledy whale statements, however authentic, in these extracts, for veritable gospel cetology. Far from it. As touching the ancient authors generally, as well as the poets here appearing, these extracts are solely valuable or entertaining, as affording a glancing bird's eye view of what has been promiscuously said, thought, fancied, and sung of Leviathan, by many nations and generations, including our own.

Herman Melville, *Moby-Dick*

Lacking ankles and feet, one thing whales do quite poorly is walk. Beached whales just lie there, never scampering about. Such was not always the case. Landlubber whales preceded blubbered whales, or so the fossils would lead us to believe. By some higgledy-piggledy process (formally known as "natural selection"), wading whales donned flippers and flukes, bulked up, and tuned their hearing to underwater sound. Some even grew baleen mustaches.

What set in motion the cetaceans' return to the sea? It's possible that families of truly ancient not-yet-whales may have submerged themselves in water to escape predation. And perhaps time spent browsing and munching in marshes or wetlands encouraged hippo-like behavior. This behavior, in turn, would promote selection of adaptations for swimming, buoyancy, and hearing destined to earn the appellation "whale."[1]

Mooshmael's fossil family photo album included several relatively recently unearthed creatures—all fine moose from his point of view and all excellent examples of whales from cousin Doris's. The definition of whales they adhere to encompasses even-toed stock (both with and without hooves) that may or may not possess flukes; ditto for flippers. In the special case of modern whales, whales are hoofless even-toed ungulates, both with and without enameled teeth. So a whale is a moose's spouting kin adapted to the life aquatic, a galloping, legless, bulky predator. Most interestingly, across all of the ancient and modern moose-whales on the Mooshmael and Doris family tree, one common trait trumps all of the rest: an expanded bulla, the bone surrounding the inner ear.

Defining a whale with consistency appears to be a confusing task. We now know that cetaceous ways of living spanning tens of millions of years have varied immensely. Cetology indeed seems unsettled—as it was in Melville's day. Recent publications, however, have resolved two major puzzles about the origin of whales: (1) as demonstrated by molecular data, whales and hippos plausibly stem from a common ancestor; and (2) as indicated by similar ankle bones, ancestral whales belong to the order Artiodactyla, the four-toed or two-toed group that has so successfully diversified into deer, cows, camels, pigs, peccaries, hippos, giraffes, and antelopes—and whales. Moose are a rather oversized type of deer and hence, as artiodactyls, might accurately claim common ancestry with whales. Hippos can make a stronger case. Skeletally, the evidence for this affinity resides in the ankle bones, which extant whales do not have. Extinct, fossilized proto-whales, however, did.

Higgledy-piggledy thinking has long plagued the gospel of cetology. The time has come for a robustly scientific definition of a whale. Whales are fluked, flippered hippos with spongy sinuses. Or maybe Melville put it better: whales are spouting fish with horizontal tails. Yet unlike other fish, they lactate.

Whale Feet on Tethyan Shores

Paleontologist Philip Gingerich and his colleagues spent the end of the 1970s digging in Pakistan for evidence of species transitions among fossil cetaceans, attempting to answer the question "Where did the whales come from? How did they enter the sea?"[2] He concluded that at first they simply walked there.

Already known were similarities between triple-cusp patterns of fossil whale teeth and fossils of a group of extinct carnivorous terrestrial mammals, the mesonychids. Fossil whales and these furry, swift fossil predators had similarly crowned teeth. For Gingerich this was a promising lead. If the tooth makes the whale, then early whales were fleet-footed and able to scamper about chasing prey in shallow water. Fish prey.

Arid mountains and cliffs above the Indus River may seem an unlikely spot to find marine and near-ocean life. But keep in mind that the top of the highest place on earth—the summit of Mount Everest—consists of fossiliferous limestone deposited on the floor of the ocean—the now defunct Tethys Sea, to be more accurate.[3] Tens of millions of years ago, when whales lacked flukes, continental landmasses were not where they are today, and a broad sea—the Tethys—separated Asia from the southern continents. The drift of India and its collision with Asia, among other movements, closed this ocean basin and thrust the Himalayas to incredible heights, modifying geography and changing the conditions of life. The ancient life of the Tethys Sea, long gone, bequeathed the great oil deposits of the Middle East and Central Asia. The India-Asia collision crumpled and lifted the sedimentary rocks formed at the bottom of the Tethys and along its shorelines. In these beds of rock lay the fossilized remains of the life adapted to ancient shores: the walking whales of Pakistan making their way in higgledy-piggledy fashion along the path to cetacean status.

Tethys-time fauna fossils abound in Pakistani rocks. Gingerich's team found fossils of both even- and odd-toed ungulates, rodents, primates, and other land animals. They discovered and named a new genus of whale based on their find of "a partial cranium with a surprisingly small braincase (even for an Eocene mammal)" and its "large lower jaw with three premolar teeth and some large sharply cusped molars belonging to a carnivorous mammal."[4] *Pakicetus*, as they called it (a combination of Pakistan, country of discovery, and *cetus*, or whale), quickly earned the title of "the oldest fossil whale." It may still stand as the smallest-brained whale ever known. For Gingerich's team, teeth and skull make a whale a whale regardless of limbs.

Where they excavated fossils, a river had once distributed its sediments into the shallow saline remnant of the Tethys. Yet the fauna chiseled from the red rocks of Chorlakki were freshwater, not marine. *Pakicetus*, as shown by its fossilized teeth and their final resting place, fed on "abundant planktivorous fishes living there."[5] This wolfish fish-eater linked the origin of whales to predatory mammals of the Eocene. *Pakicetus* fulfilled Gingerich's quest for a transitional form between whales and land mammals. Its behavior pointed toward the evolutionary opportunities of a seagoing way of life. It even appeared to have some of the bone thickening characteristic of the internal whale ear, essential to detecting sound underwater.

Bodiless head fragments made at best a tenuous link between whales and land mammals, but they suggested where to look and what to look for. For a time, cranium, jaw, and teeth appeared to link *Pakicetus* to the primitive Asian family of mammals, the mesonychids, comparable in their time to wolves and hyenas of today. About 60 million years ago the mesonychid *Sinonyx* walked somewhat flat-footed as it prowled near streams. Its long, narrow snout held molar teeth suitable for slicing through the flesh of its prey, which most likely included fish.[6]

Pakicetus turned out to share key traits with the older *Indohyus*, a member of the family Raoellidae in the order Artiodactyla. Molecular data and the double-pulley shape of the astragalus bone in the ankle of *Indohyus*—an attribute characteristic of artiodactyls—clinched the reclassification and ushered the mesonychids off the auditioning stage for cetacean ancestry status. The structure of the astragalus is in part the key to the powerful flex of the artiodactyl's hind leg and its double-pulley design (rounded ends) is a unique trait that defines membership in this group in ancient or modern times. Positioned between the end of the tibia (shinbone) and the bones of the foot, it hinges the ankle.[7]

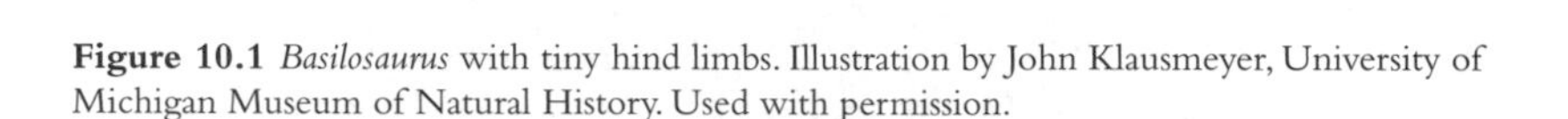

Figure 10.1 *Basilosaurus* with tiny hind limbs. Illustration by John Klausmeyer, University of Michigan Museum of Natural History. Used with permission.

Modern ungulates are hoofed vegetarians with reduced toe counts. Sixty-five million years ago, proto-ungulates could count among their kind eaters of almost everything. With *Pakicetus* classified as whale-related and granted, along with *Indohyus*, artiodactyl membership, moose and whale have good reason to claim common, but very distant, ancestry. Recall that Mooshmael recognized Very Very Great Uncle Pakky as his ancestor.

Skeletal elements of several other pakicetid creatures have since joined the inaugural collection of archaic whale fossils bequeathed to paleontology by Gingerich and his colleagues. They reinforce the link between ancient artiodactyls and modern cetaceans. True, raoellids were raoellid artiodactyls, not cetaceans (or archaic cetaceans). Eventually Gingerich proved the existence of *whale feet* in the fossil record of archaic whales.

Again he was probing the shores of the ancient Tethys, this time in Egypt's Valley of the Whales. The team unearthed hind limbs tiny in size compared to their bearer, a sixty-foot-long serpentine whale from 36 million years ago known as *Basilosaurus isis*. The fossil revealed whale feet to the world for the first time. "Feet" is a misnomer if function is used to classify the bones rather than anatomical position and bone structure. *Basilosaurus* probably used its tiny hind feet as claspers to aid in copulation. This creature, silly-looking enough with its tiny hind limbs, made talk about walking whales seem much less preposterous. *Basilosaurus* resembled a morbidly obese dachshund with the snout of a crocodile—an overstuffed sausage with ridiculously small hind oar-limbs. Its nostrils had not migrated to the blowhole position at the top of the head. Mooshmael's Very Great Uncle Basil was a *Basilosaurus*—a serpentine whale of a moose, minus antlers.[8]

A Whale of a Hippo

One might say whales are exactly the same as hippos except for the differences (or, equivalently, they are completely unalike except for the similarities). In the jargon of evolutionary biology, both are "whippomorphs," bequeathed as sibling descendants with features of their common ancestor, the first whippomorphid.[9]

One less than obvious difference that distinguishes them is the bone structure that envelops the middle and inner ear. This bone—the bulla—thickens in whales, and thickening was present in the raoellid *Indohyus*. Branching from a common ancestor, cetacean hearing went in one direction and hippopotamid in another. Hippos (a four-toed artiodactyl) lack a thickened bulla, an enhancement that their whale cousins perfected for underwater listening.[10]

Biochemical evidence has further resolved the branching pedigree of whales and their hippo kinship. Testing for antibody reactions among living whales yielded reactions similar to those of particular artiodactyls.[11] It was as if whales mysteriously shared allergic responses to the same allergens as deer and antelope, suggesting a persistent family trait. Whales may have evolved these biochemical affinities independently, but the antibody data do not stand alone. DNA sampling has reinforced the inference of whale-artiodactyl common ancestry.

For example, geneticists have tested artiodactyl and whale DNA sequences for similarities in regions referred to as "non-coding." Randomly accumulating genetic mutations in sections of DNA that do not code for essential proteins often provide the raw material for tracing common ancestry. More time means more mutations in non-coding regions. By keeping quiet, these mutations escape natural selection's watchful eye. Fewer differences in corresponding strands of DNA thus mean closer relationship.

In *Darwin's Ghost*, author Steve Jones offers a lucid summary of the genetic research that has helped to puzzle out whale ancestry. He explains how non-coding data ally whales closest to hippos within the artiodactyls and closer to giraffes and deer than to the rest of the cloven clan. Whales, hippos, giraffes, and deer share three sequences of DNA inserted into the genome of a common ancestor long ago by a retroviral infection. Relatively harmless and inert, these sequences enjoyed a free ride from generation to generation, infecting all of the ancestor's progeny. These sequences are missing from all other artiodactyls, including pigs and camels.[12] Deer, and therefore moose, by such an analysis are more closely related to whales and hippos than they are to pigs and camels.[13]

There is reason to assume that conclusions derived from DNA sequencing might not hold: mutation reversals can erase differences and fool the researcher into believing in a closer relationship than truly exists. Sampling more regions of the genome is the way out of this conundrum.

Coding regions of the genome provide relevant data as well non-coding ones. Lactation, a defining attribute of mammals, produces milk, and milk has protein coded for in DNA. Comparisons of segments of DNA coding for casein proteins in milk from cetaceans and other placental mammals reveal a closer relationship—better matching sequences of DNA—between hippos and whales than between whales and any other artiodactyl group.[14] The story of shared ancestry told in casein proteins is the same as written by viral insertions, reactions to allergens, and the shape of ankle bones.

The initial fossil evidence encouraged paleontologists to entertain a mesonychid origin for whales. Ultimately, the ankle bones simply said "no" and the molecular data convincingly established hippos as the closest living relatives of whales. Fossils are scattered and few, molecular data perilous in their own ways. Yet when molecular and fossil evidence shake hands firmly, the conclusion sticks. Whales descend from primitive artiodactyls most closely related to modern hippos. From that point on in the gospel of cetology, what makes a whale a whale—and not a fluked hippo with a bulky bulla—is a matter of human judgment.

A Spouting Fish with a Horizontal Tale

The time is 1818; the setting: Mayor's Court, City of New York. City Recorder Richard Riker presiding as judge; expert testimony being presented by Captain Preserved Fish, respected whaleman of New Bedford, Massachusetts, on behalf of the defendant, Samuel Judd. Without reservation, Captain Fish declares whales, along with porpoises, to be air-breathing mammals. Unlike fish, they have tails that spread horizontally, and their fins are anatomically akin to "arms." The encyclopedia says so. In cross-examination, Captain Fish is unable to identify bones in the fin of a whale corresponding to those of the human arm. As a result, he helps the defendant very little, and momentum shifts to the prosecution.

Another experienced whaleman, James Reeves, takes the stand. Whalemen, he explains, indeed hunt "fish." They speak to one another of the number of fish captured or held fast to the ship. Of course, he acknowledges, whales breathe air. As to water-breathing, typical of fish, James Reeves believes no one has proven whales completely unable to breathe while submersed.

The jury deliberates and reaches a verdict in fifteen minutes. At stake is more than Samuel Judd's innocence or guilt. Captain Fish's testimony bears on whether or not whales are fish, and whether or not whales are fish determines whether fish oil tax is due on whale oil. A verdict must favor either a philosophical, scientific interpretation of "whaleness" or common knowledge of "fishness" anchored in religious understanding, but not both. The idea of Darwinian evolution, yet unborn, is not on trial, but its antecedents are.

D. Graham Burnett's *Trying Leviathan* completely captivates the reader with the history of this early-nineteenth-century legal tussle enmeshed in biological nomenclature.[15] His exquisite re-creation of the trial and its characters reveals the forgotten historical significance of the question "Is a whale a fish?" From the comical name of Captain Preserved Fish (historically accurate) to the debate over what would now seem a well-settled matter, Burnett's story of conflict between scholarly opinion and common sense unfolds as tax collectors and commercial businesses jostle to affect the outcome. Then as now, oil held center stage in business and commerce. Its procurement, regulation, and taxation mattered greatly to many, the same as today. Interested parties from worker to researcher, from government official to businessperson, awaited the verdict. What was at stake? An oil tax.

In question was not just any oil tax but a particular tax on barrels of fish oil sold in New York State. "The well-known candle maker and oil merchant, Samuel Judd, had refused to pay the inspector's fee on three casks of spermaceti oil—protesting that the oil was not 'fish oil' but 'whale oil' and that whales were not, in fact, fish."[16] The tax collector depended on this income from measuring, gauging, and inspecting casks of fish oil. Judd refused to pay on the sound logic that since whales were not fish, whale oil was not fish oil. The question of his evasion of a tax turned on the definition of "whale." "Only fish oil is fish oil," Leo Lionni's minnow would no doubt argue.

Maurice v. Judd ended in triumph for the plaintiff. The jury, in short order, found that a whale is indeed a fish and that therefore whale oil is certainly fish oil. The tax was past due. Commonsense perception—if it swims in the sea, it's a fish—and the expert voices of whalemen prevailed. The jury dismissed the views of elite "natural philosophers" (scientists), who based their reasoning on breakthroughs in the study of anatomy, as simply irrelevant. How an animal lived, and therefore how the animal interacted with commercial fishing interests, determined its classification, not the esoteric details of its inner anatomy such as limb structure or its metabolic processes such as air-breathing. And besides, no one could know with certainty just how whales breathed when submerged. Whaling was a bona fide fishery, its catch the "royal fish" of kings. Whale fish was whale fish.

Figure 10.2 Casks of fishy whale oil on trial. Illustration by Jan Glenn.

 Many citizens today would find *Maurice v. Judd* inexplicable and ridiculous. Whales are most certainly not fish. The features scientifically defining fish, bird, amphibian, reptile, and mammal have saturated the public. They are inculcated by every visit to a natural history museum.

Nevertheless, resolving the initial contradiction (a whale is both a fish and not a fish) faces little difficulty once the competing categorization systems have been aligned with their corresponding purposes. A whale is a fish when categorized by its external features (horizontal "fins"), behaviors (swimming), and habitat (sea). A whale is not a fish—it is a mammal—when categorized by its metabolism (warm-blooded), behaviors (nursing), and ancestry (descent from land animals). The purpose pursued determines which set of criteria to choose. Once the purpose is clear, a scheme for classifying may be judged either useful or inappropriate. Either way of categorizing a whale—fish or non-fish—serves a purpose. And successfully accomplishing the purpose validates the choice of the categorization scheme. The difficult problem is achieving shared purpose.

For the biologist (and a host of premed students) the shared purpose is comparing anatomies and drawing inferences about body mechanics as well as evolutionary affinities. Structures and their functions account for the ability of the whale to swim using vertical undulations of the fluke. Swimming is nearly incidental to the classification of the whale, not truly essential.

Almost certainly, Herman Melville was well versed in the case of the casks of fish oil. Melville's taxonomy of whales fitted the whaler's need: to know where to hunt which kinds of whales, what dangers to expect, and what value to ascribe to their fats and oils. If any of Ahab's crew had been called into court, giving testimony under oath would have meant sharing without hesitation the self-evident proposition that a whale is a fish, therefore that whale oil is fish oil. Melville's nineteenth-century readership would likely have sympathized with the verdict in *Maurice v. Judd*—the finding that whales are perfectly fine fish. Swimming was what mattered most; and whales, like fish, swim expertly. Melville faulted the Linnaean classification of whales as mammals because it drew upon irrelevant (to him) observations of their movable eyelids, hollow ears, a warm two-chambered heart, and *penem intrantem feminam mammis lactantem*, translated from the Latin by *Moby-Dick's* editor Charles Feidelson Jr. as "a penis which enters the female, who suckles by means of teats."[17] These internal aspects failed to bear on the taxonomic status of whales to whalers.

Melville's protagonist, Ishmael, paused to consider, "Now, how shall we define the whale, by his obvious externals, so as conspicuously to label him for all time to come? To be short, then, a whale is *a spouting fish with a horizontal tail*."[18] As logical and accurate as it was, the label has not served well for all time to come.

No doubt Lionni's minnow would agree with classifying a whale as "a spouting fish with a horizontal tail." To the minnow, his tadpole friend was a leg-sprouting fish with a vertical tail. The minnow thought in terms of externals—fins and limbs—the same as Ishmael.

Ishmael next commented on the taxonomic status of marine animals similar to whales. "A walrus spouts much like a whale, but is not a fish, because he is amphibious." Ishmael faulted naturalists who included dugongs and manatees ("pig-fish" and "sow-fish" to his mariner kin) among the whales. "But as these pig-fish are a nosy, contemptible set, mostly lurking in the mouths of rivers, and feeding on wet hay, and especially as they do not spout, I deny their credentials as whales; and have presented them with their passports to quit the Kingdom of Cetology."[19] Failure to spout earned dugongs and manatees banishment from Moby Dick's realm.

Ironically, evolutionary biologists now connect dugongs and manatees through common descent more closely to the elephants than to the whales. Melville had a point.

From Ishmael's perspective, dugongs and manatees lacked something more significant than spouting. The preeminent feature of being a whale was its majestic, almost supernatural aspect, its embodiment of power stemming from an ineluctable union of malevolence and indifference. The value he ascribed to whales impelled him to separate other marine mammals from this class in much the same fashion that esteem for human life compels many people to balk at classifying themselves as animals—as mere *two-legged fish that breathe air, spout ideas, and lack tails.*

In the modern world comparative anatomy and genetics rule the rules that name the natural order, and scientific authority does not defer to the expert opinion of hunters of fish. Whales are not fish, and anyone who thinks they are is guilty, in the court of science, of holding a basic misconception.

Samuel Judd, defendant in the fish oil trial, invoked the testimony of natural philosophers who placed whales in the context of tetrapod anatomy (even though they lacked hind limbs) and examined their parts in relation to the anatomical structures of mammals. James Maurice, the inspector who took Judd to court, introduced the testimony of whalemen who placed whales in the context of sea commerce and their oil in relation to other products from the sea.

Nature creates no names; people do. They invent categories to suit their purposes. Therefore, categorization schemes vary as purposes differ. Taxing oils collected at sea and handled in port is one purpose; arranging animals in groups according to shared anatomical traits is another. In 1818 natural philosophers, following Linnaeus, believed that shared anatomy revealed the natural order of Creation. A half century later, Darwin's disciples demonstrated that such natural ordering revealed instead a history of shared ancestry.

Labeling a whale a fish is not wrong, just the consequence of categorizing according to a purpose not shared among evolutionary biologists. Such impertinence!

Naming the nose inherits some of the same difficulties as deciding what makes a whale a whale. Perhaps no more striking example exists of the diversity of mammalian structure based on a shared trait than the contrast between an elephant's trunk and a whale's blowhole. They are two very different ways of positioning nostrils.

A whale, especially a sperm whale, is a gigantic head and mouth with just enough body left over to contain a stomach and attach a fluke. The mouth is big and the head massive. But where is the nose? Claiming "that which stands between two eyes and north of a mouth is a 'nose,' and that is that" will not do.[20] Naming the front of a whale a mere "nose" is an affront to its dignity.

Where is the majesty, horror, and irrepressible malevolent Will in the image of Ahab's nemesis striking a whaling vessel *with its nose*? "Nose" sounds too silly and too mundane to refer to the massive brow of the White Whale, a brow bespeaking inscrutable wisdom, overwhelming power, and fearless contempt. No; classifying that which stands between the eyes of a whale and north of its mouth as a nose will not do because it leads us to chuckle at the thought that Moby Dick slammed his *nose* into the *Pequod*.

The front end of the sperm whale is mostly a boneless mass, wrapped as toughly as if "paved with horses' hoofs." It is a "dead, blind, wall."[21] There is nothing external; no nose, if external prominence is essential to its definition. What counts as a nose, its spout hole, sits atop its head.

Or maybe the whale is mostly nose, and hard-nosed at that. Unless, of course, the front end of this whale is a stiff upper lip, the nose pushed to the top of the head. Melville clarifies the situation: "Physiognomically regarded, the Sperm Whale is an anomalous creature. He has no proper nose. . . . A nose to the whale would have been impertinent."[22] A whale is much too serious an emblem of the nature of Nature to have something as undignified as a wiggly nose.

As Melville noted, experts in his day spouted a great number of higgledy-piggledy statements about whales not to be taken as cetacean gospel. The story of whale ancestry can be told, however, while leaving the definition of a whale undecided. It includes a cast of characters transformed from fishing raccoons into krill-swilling beasts, transmogrifying mammalian nostrils to an unprecedented degree. The sinister snout gave way to the furrowed brow; growls softened to become the humpback's song.

Mooshmael's Family Tree

In 2007, a team of renowned fossil whale experts led by J. G. M. Thewissen published the telling research article, "Whales Originated from Aquatic Artiodactyls

Figure 10.3 *Indohyus*, an Eocene creature from India and a likely progenitor of whales, as reported by Thewissen, Cooper, Clementz, and Bajpai in 2007. Illustration by Jan Glenn.

in the Eocene Epoch of India."[23] "Eocene" refers to ages on the order of tens of millions of years ago (34 to 56 million years ago, to be a bit more precise), during the early heyday of the "Age of Mammals." By 53.5 million years ago, artiodactyls constituted a diverse group poised to spin off the cetaceans. As in Pakistan, Eocene sedimentary rocks from India yielded provocative clues about the origins of whales—the skeletal remains of *Indohyus*.

In their reconstruction, the even-toed *Indohyus* had the limbs of a capable wader. Thewissen argued that whales diversified from such artiodactylic ancestors first through wading adaptations to aquatic habitats and later by exploiting opportunities to prey on fish in open waters. Ultimately, these creatures' descendants bulked up and began to strain the ocean for food in massive quantities.

Bulk was unimportant to *Indohyus*. The little wader was less than one meter long from the tip of its nose to the tip of its tail. Its whale-toothed head was relatively large for its body, but still only about ten centimeters in length. This creature was Antie Indo of India to Mooshmael. A dainty fawn to Mooshmael's grand size, she, like Mooshmael, sported a flexible ankle bone with pulley-shaped ends. At least Mooshmael could relate to the wading lifestyle she practiced among the lily pads of her epoch.

The last 50 million years of earth history have borne witness to mind-bending examples of additional candidates for whale ancestry. Fossils provide tantalizing glimpses of the higgledy-piggledy transition from legged to flippered whales and from tailed to fluked swimmers. Giant whale pinky fingers are no longer with us, and whales, dolphins, and porpoises have no hind limbs, of course. They do possess flukes that pivot from the lower back and flippers that wed together the bones of the forelimb into stiff stabilizing and steering fins. Achieving this efficient form left a host of magnificent whales-not-yet-whales in its wake.

Say "Bigfoot" and the image of a whale seldom comes to mind. That's because all of the big-footed whales are extinct—as are all the whales with little feet for that matter. At about 50 million years ago, *Ambulocetus natans*, whose name means "walking whale that swims," propelled itself flukeless through the water by paddling its forelimbs and wiggling its tail in otter-like fashion.[24] Its discoverers described it as walrus-sized with convexly curving toes seventeen centimeters long—a whale of a toe.[25]

Turning a Bigfoot whale into Moby Dick whose flick of a fluke could capsize whaleboats and power its flight across the breadth of the Pacific would seem improbable. Nevertheless, on the heels of *Ambulocetus* by a few million years swam archaic whales much less otter like. As inferred from fossil vertebrae, *Artiocetus clavis* flexed and extended its spinal muscles so as to move its hindquarters vertically. The up-and-down motion produced the thrust it needed. Its propulsion was fluke-like, without the fluke, and its forelimbs stabilized and steered this proto-leviathan. Its nose opened well forward on its face. Probably an active wader, *Artiocetus* boasted powerful triceps muscles and a flexible elbow joint. Its hand digits spread widely and its "pinky" was nearly as long as its forearm. There was no blowhole nor whale brow to contemplate in this creature. Its name captures two groups: *artiodactyls*, hoofed creatures, and *cetaceans*, or proper whales.[26]

Dating from 46 million years ago, *Rodhocetus balochistanensis's* (Mooshmael's Very Great Aunt Rodhie) skeletal remains, similar to those of *Artiocetus*, included a flattened femur, less cylindrical than in its cousin species and further evocative of paddling. Its forelimbs ended in multiple digits supporting a generalized mammalian wrist. On land it walked on the webbed digits of its hands and probably wobbled along much like a sea lion.[27]

Rodhocetus could spread webbed toes on its hind limbs wide to make powerful swimming strokes, then contract and narrow its foot when gliding. It may have folded its forelimbs against its body and probably depended upon tail motion for propulsion when submerged. *Rodhocetus* could take strokes with its front legs as well. Given its large hind feet and rudder-like tail, an otter in fluid motion is the best living analogue for its swimming behavior.[28]

Rodhocetus was a flukeless whale propelled by paddle-form limbs. Its long snout pushed its nostrils far forward of its eyes. Its locomotion on land may have resembled that of an exceptionally agile pinniped. The beast could dart through the water and give chase up the shore, a killer whale on land and at sea.

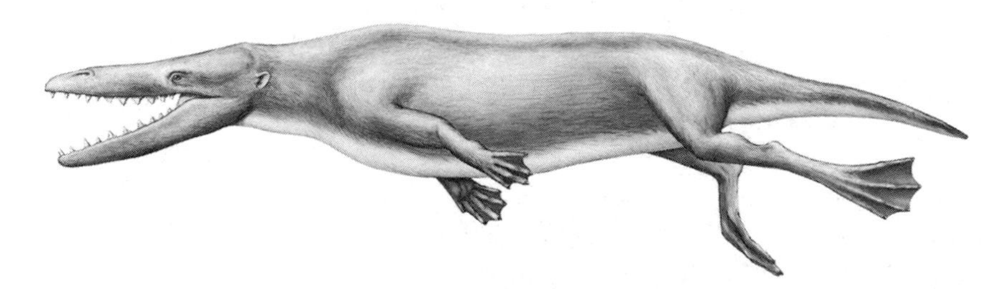

Figure 10.4 *Rodhocetus*, a paddling proto-whale. Illustration by John Klausmeyer, University of Michigan Museum of Natural History. Used with permission.

Recall for a moment the Tethys Sea's *Basilosaurus* recovered from Egyptian deposits. The discoveries of *Indohyus* and *Basilosaurus*, together with a cetacean cast of many other characters, freeze the frames of an incredible evolutionary journey at important junctures millions of years apart. No one can resurrect *Basilosaurus* of tiny hind limbs fame, provide it with mates, manipulate the globe geologically and climatically to make it the same as it was 30 million years ago, and see if a modern whale happens. Nor will such a procedure fare any better with a *Pakicetus* or *Indohyus*. Such expectations of the experimental method are absurd; puzzling out descent calls for a different method of inquiry for which samplings of fossils, despite their rarity, work quite well.

The principal method of inferring relationships, and hence degree of common ancestry, among a set of creatures reduces to a single word: homology. The term means essentially "inheritance of the same parts, in the same places, with some modifications." The convention in evolutionary biology is to infer common ancestry from the degree to which different organisms share inherited, homologous traits. Homologies may describe features at the anatomical or molecular level. Inferring descent or at least degree of relationship depends on recognizing homologies among inherited traits. In Darwinian science the debates about kinship shift back and forth between different kinds of homologies: skull, inner ear, limbs, organs, DNA.

Limb homologies have played a vital role in solving the puzzle of the origin of whales. Two key ankle bones found in the remains of *Artiocetus* and *Rodhocetus*—the cuboid and astragalus—convinced Gingerich and the paleontological community that whale ancestors branched from the artiodactyls quite directly, not from the wolfish mesonychids of the Tethyan shore, as they had hypothesized upon their earlier finding of *Pakicetus*.[29]

Whales no longer have hind limbs. Higgledy-piggledy they disappeared. Homologous traits among intermediate forms—in this instance, *Rodhocetus*—are the key to linking living species to their ancestral populations. Multiple homologues whose ranges in time clearly overlap convincingly connect the endpoints.

Intermediate or "transitional whales" represented by Very Great Aunt Rodhie have vertically flexible backbones and skulls with inner ears embedded in special chambers. These homologies strongly suggest common ancestry with modern whales and thus link Aunt Rodhie forward to Uncle Basil. Aunt Rodhie's limbs link her backwards in time to artiodactyl stock. More ancient creatures, the Auntie Indo type, have limb and snout structures that demonstrate common ancestry with extinct ungulates.

Very Great Aunt Rodhie certainly took her transitional role quite seriously. Forward in time from Aunt Rodhie's day, whale limbs atrophied and fluke-like flattening displaced the otterish tail, yielding basilosaurids—beasts with little hind limbs, as silly-looking on a whale as the puny forearms on the mighty *Tyrannosaurus rex*.

Evolution appears to have pulled the ancestors of the whale by the ear into the sea. The mammalian ear features a "tympanic bulla." This is a bone that surrounds the inner ear, home to detecting balance. In whales, this bone thickens and functions to transmit sound effectively underwater. *Pakicetus inachus*, Gingerich's oldest (approximately 56 million years ago) whale ancestor candidate, had a thickened bulla but not necessarily as an aid to underwater hearing. *Pakicetus* had a mammal-like eardrum that could not work in water. Why the thickened bulla? The better to pick up vibrations through the ground, as do turtles.[30]

Indohyus indirae, Thewissen's raccoonish aquatic Eocene wader, also had a thickened bulla. Observations of skull, tooth, and skeletal features of these small land creatures suggest affinities with cetaceans.[31] Traits trending toward whaledom, such as a thickened bulla, functioned in tandem with limbs quite well in the nearshore and aquatic wading life of Eocene habitats. The thickened bulla and a snout suited for fish-grabbing opened a path to the sea for the descendants of *Pakicetus* and *Indohyus*.

Thewissen believes that *Ambulocetus* had the beginnings of a channel linking its jaw to its ear that enhanced the detection of ground vibrations. By resting its jaw on the ground, a strategy seen in modern crocodiles, *Ambulocetus* could have listened for approaching prey.[32] The croc's jaw and ear play off each other, picking up good vibrations. Hearing through solid ground perhaps pre-adapted whales to hearing underwater, thus providing the structures (thickened bone) for evolution to exploit.

By the time of the first fully aquatic whales, some 40 million years ago, ear canals had closed and thick tympanic membranes (eardrums) had formed. From then forward, whales profited from a splendid arrangement around the middle ear of air sinuses filled with sponge-like tissue. This tuning permitted whales to resolve sounds rather precisely and even determine their direction, for the sinuses isolated sounds received from the left and right ears. Whales even evolved a fat-filled canal capable of conducting sound to the middle ear from the lower jaw, the structure foreshadowed in *Ambulocetus*.[33]

Hearing is but one function of the ear. It responds to pressure and hence is crucial to balance, orientation, and movement—as anyone who has spun around too many times can testify. It is an inertial guidance system that detects acceleration up or down, left or right. It tracks turning. Fossils indicate that whale-like traits of the inner ear related to spatial orientation were well in place before oceanic hearing mechanisms evolved.

Hearing adaptations accompanied the derivation of the fluke from the tail, an opportunistic exploitation of a vertically flexing backbone. As the fluke reduced the locomotive utility of the hind limbs, natural selection fashioned the forelimbs into flippers, all in about 15 million years. By that time, the higgledy-piggledy whale verged on modernity.

Toothed whales (odontocetes) can echolocate prey via high-frequency sound; baleen whales (mysticetes) are able to communicate with one another over vast distances via low-frequency sound. The 28 million year old fossil skull of a toothed whale, *Cotylocara macei*, discovered by Jonathan H. Geisler, Matthew W. Colbert, and James L. Carew, displays several traits strongly associated with echolocation. Among these are bony adaptations such as an asymmetrical cranium and broad upper jawbones.[34] Presumably, echolocation among toothed whales dates back to the Oligocene epoch. Fossils from 34 million years ago, the oldest examples of mysticetes, are less conclusive about the hearing capabilities of baleen whales.[35]

Back to Baleen

Kipling wondered how the whale got its throat—the baleen used in filter-feeding and capable of supporting immense bulk. "Suspenders and a shredded raft left by a mariner" was what he proposed for its origin. It's time for a more serious proposal.

Flexible yet stiff, baleen is made of the protein keratin, not raft rubber, and grows from the palate. There is no homologous relationship between baleen and teeth; baleen is not a modified set of teeth. The origin of baleen is simultaneously an origins problem of tooth loss, for the baleen whales, as even Darwin knew, form teeth while in a fetal stage of development. These teeth are resorbed before the whale reaches adulthood.

The genetic instructions for growing tooth enamel vary among living whales. That variation, plus the presence of teeth during early development among baleen whales, provides evidence that there once existed an intermediate whale sporting both teeth and baleen.[36] Teeth, presumably, were present even among the ancestors of baleen whales, and a fossil whale has stepped forward to make this point. The fossilized palate of the adult whale *Aetiocetus weltonni* features structures for supporting teeth and nourishing baleen (preservation of baleen itself is rare and inferred for aetiocetids). Dating from between 24 and 28 million years ago, it did not reabsorb its enameled teeth during development.[37] The date suggests a time when lineages leading to sperm (toothed) and blue (baleened) whales had the potential to diverge.

Individuals vary from one to another; some survive, others do not. As a result, populations—not individuals—adapt to the changing conditions of life. Individual by individual, thought Darwin, selection preserves and discards variants. Any observed trend—fins to limbs, limbs to flippers, tails to flukes, gums to bristles, bristles to baleen—is a function of hindsight, not a predestined path.

Surviving is a chancy business where uncertainty reigns, serendipity matters, and opportunism triumphs. All forms of life are constrained by their past yet have unpredictable potential for the future. Buffered by chance mutations, an organism gambles on survival by betting its inheritance. Sometimes it gets lucky and squeezes through circumstances quite unlike those survived by past generations. Being what a whale is today, given what a whale has been, is the best bet on getting by as a whale tomorrow, but it's not a guarantee. Yet what whales may become cannot be known.

The majesty of cetacean evolution encompasses the most remarkable of transitions among beasts. It is an epochal story of Melvillian proportions as creatures walked, paddled, and galloped into the sea. The journey has bequeathed the planet with behemoths of unknown sagacity able to listen across thousands of kilometers of ocean emptiness or locate objects in the murkiest depths.

The evidence from fossilized forms, molecular analysis, and species diversity tells what a whale is and has been. Predation looked seaward; limbs were lost; flukes flourished; heads and sinuses enlarged; teeth came and went; ears adapted to living in the sea; baleen mustaches burst forth; bulk increased: in a "higgledy-piggledy" fashion, leviathans happened.

11

ARCHAIC CHICKENGATORS

Amniotic Archosaur Ancestors of Dinosaurs and Birds

> *A few days later, the frogs heard a strange noise coming from the egg. They watched in amazement as the egg cracked and out crawled a long, scaly creature that walked on four legs.*
>
> *"See!" exclaimed Marilyn. "I was right! It is a chicken!"*
>
> *"A chicken!" they all shouted.*
>
> Leo Lionni, *An Extraordinary Egg*

Marilyn, Jessica, and August, the three frog companions in Leo Lionni's story *An Extraordinary Egg*, had found a new friend, the just-hatched chicken. It was an unusual chicken to be sure. Covered in scales, it had no feathers. The strange little chicken could swim with ease. It paddled with four legs and a long tail. It could not fly. But it hatched from a chicken egg, or so they believed, and was therefore a chicken. (Frogs are quite capable of using "moose logic.")

The frogs had a wonderful time playing all day long in the pond with their new friend, the scaly chicken, who taught them "new ways to float and paddle." After many days, Jessica and the chicken learned from a red and white bird that the chicken's mother was nearby. The frog and the chicken went in search of the mother chicken. It did not take long before "they came upon the most extraordinary creature they had ever seen. . . . It was asleep. But when it heard the little chicken shout 'Mother!' it slowly opened one eye, smiled an enormous smile, and, in a voice as gentle as the whispering grass, said, 'Come here, my sweet little alligator.'"[1]

Jessica, Marilyn, and August had misidentified the egg. They were, after all, just frogs, with eyes adapted to spotting flies flying. Frogs lay eggs in watery places, not on dry land. Chickens lay eggs on land. Land eggs have tough coats and look much the same as large pebbles—at least to pebble-collecting frogs such as Jessica, Marilyn, and August. The reader, however, can tell at once from Lionni's illustrations that the hatchling was no chicken. It was, of course, a baby alligator.

Amphibians, reptiles, monotremes, and birds—frog, alligator, platypus, and chicken—all hatch from eggs and all boast backbones. Yet the eggs of amphibians, and many of their traits, differ from the others' in striking ways. Lionni's story hides some fascinating evolutionary twists linking these creatures together while defining their differences. In between the lines lurks the incipient origin of earth's most monstrous creatures: the dinosaurs.

Telling this other story of egg-stamped origins does introduce intimidating jargon and multisyllabic names. The terms, however, are useful in tracking descent. Approached with whimsy—with thoughts of easily deceived frogs—the jargon entertains as well as guides thinking. Moreover, descriptions of eggs and skeletons may serve as primers for a visit to any Age of Reptiles gallery at a fine natural history museum. Alien and forgotten among many adults, dino-speak charms ten-year-olds. To best enjoy the jargon, repeat the words out loud often and teach them to a child.

Archosaur Ancestry

Amphibious tetrapods came ashore in the Devonian period, on the order of 360 million years ago. The air proved dangerous. It dried things out, especially eggs. Not to mention skin. To dominate the land, vertebrate life had to adapt.

In time, a group known as archosaurs ("ruling reptiles") solved the problems of desiccation quite expertly, as did other reptilian beasts that were roaming across the land by the dawn of the Permian period, some 300 million years ago. Archosaurs appeared amidst the parade in the Middle Permian. Before the birds, before the dinosaurs, and well after the origin of reptiles, as the arid Permian came to a close 250 million years ago, archosaurs began their march to prominence in the Triassic. They split from their reptilian kin—the turtles, lizards, and snakes.

In March 2015, discovery of a fearsome predatory archosaur captured news headlines.[2] Approximately 251 million years ago, *Carnufex carolinensis*, the "Carolina Butcher," stalked the supercontinent Pangea, then in the early throes of breakup. The Carolina Butcher was a quite large—now here begins the fun-to-say jargon—crocodylomorph. Like a grizzly bear with scales, it could rear up on its hind legs. It was nearly three meters of croc with a half-meter-long skull packed with scissor-blade teeth.

Today's birds and crocodiles—the chicken and the alligator among them—are the living descendants of ancient archosaurs. They boast a pedigree in common with the Carolina Butcher. Even the flying reptiles (pterosaurs) came from archosaur stock. So too did the dinosaurs who took center stage in Late Triassic times (220 million years ago). Birds followed by taking wing over 150 million years ago, late in the Jurassic.

Eggshells and body scales—and very efficient kidneys for retaining water—contributed to archosaur and archosaur progeny success. Dinosaurs, then birds, perpetuated the archosaurian approach to living on land, where conditions made running and flying good options. Tucking the legs under the body for efficient locomotion, an adaptation stumbled upon by one group of archosaurs, enhanced these behaviors.[3] Perhaps *Carnufex carolinensis* should be credited with contributing to this stumble.

Archosaurs sprawled and archosaurs crawled and archosaurs small had the speediest limbs of them all. Long-snouted, they socketed their teeth. A hole in the skull between the eye and the nose marked their kind. Archosaurs were lifelong air-breathers that tore through shells and hatched from eggs, not gilled tadpole babies that wiggled to escape from gelatinous blobs. Nearly all of the archosaurian dinosaur progeny became extinct. "Nearly all" implies that some dinosaurs did *not* become extinct. The name for non-extinct dinosaurs is a simple one: birds.

Archosaurs and amphibians, two distinct forms of tetrapodding, have coexisted for hundreds of millions of years. They do, of course, share a number of skeletal features that emerged as backboned life departed the sea. Nevertheless, different ways of swallowing and breathing plus distinct skull and limb anatomies—and the aforementioned eggshells and scales—set these groups apart. For frogs such as Jessica, Marilyn, and August, archosaurs were stereotypical "others." All archosaurs must look pretty much alike to the average amphibian.

Membrane Links

Clearly, the hatchling archosaurian alligator seemed very birdlike to Lionni's frog friends. Croc, alligator, chicken, duck—all variations on the same sort of critter to them, and all hatched from pebbly eggs, whether leathery or hard-shelled. The frogs had difficulty distinguishing an alligator from a chicken, as it turns out, for good reason.

The hard, pebblish egg that so intrigued the three frog friends had membranes that frog eggs lack. Beneath the shell of bird, crocodile, lizard, snake, turtle, and tortoise eggs—not to mention archosaur eggs—a membrane called the amnion surrounds the developing embryo. Covered by a calcium-rich eggshell secreted by the mother, the amnion contributed to the opportunity for backboned life to exploit terrestrial habitats.

So Jessica, Marilyn, and August recognized a pattern that had endured for a great while among terrestrial inhabitants: the amniotic egg—the egg that looked to them like a pebble.

There's more to an amniotic egg than shell, yolk and white. The egg is a wonderful life raft on land. Its membranes store food, prevent desiccation, and permit gas exchange. The amniotic bag of seawater provides an embryological nursery: a balloon world of security. Outside the amnion are the chorion, which adds protection and exchanges gases; the allantois, which collects wastes; and, of course, the yolk sac, containing nourishment.

All animals that begin life within the amnion are called "amniotes."[4] Some hatch from eggs; some are born live. All mammals (and therefore humans) enclose their developing embryos within the fluid-filled amniotic sac, as do turtles, birds, snakes, and alligators. Mammalian amniotes branched apart from other amniotes 60 million years prior to the ascendency of the archosaurs. Whether hatchling or live-birthed, all beneficiaries of an amniotic membrane are descendants of some primordial amniotic creature, an even more inclusive club than the archosaurid one. Archosaurs are the non-mammalian, non-snake, non-turtle, non-lizard amniotes. Crocodiles and alligators are their most enduring, if not most endearing, primitive forms.

All amniotes are vertebrates, but not all vertebrates are amniotes. Amniotes live both on land and in water and all have backbones. Amphibians live on land and in water and also have backbones, but not an amnion. Early on, frogs and salamanders found themselves on the non-amniotic branch of the vertebrate tree of life. Club Amnion has always excluded the amphibians.

Walruses, killer whales, marine iguanas, garter snakes, box turtles, duckbill platypuses, bald eagles, and people all begin their lives as embryos snug in an amniotic world that mimics the sea. Dinosaur embryos were no different. Dinosaurs, well-respected running and walking specialists, are extinct terrestrial, vertebrate, reptilian animals that laid eggs with amniotic membranes. Membrane tissue does not preserve well in fossils, but the inference that dinosaur embryos developed within an amnion is strong and noncontroversial, gaining them membership to the club.

The Crocodile's Smile

Just as sharing an amnion during embryonic development defines a group of related animals, so too do other prominent inherited traits define subgroups within the amniotes. Holey patterns in the skull often indicate kinship. Skin and scales hid this pattern from Jessica, Marilyn, and August's view, of course. To them only the egg and its hatchling boasted traits of archosaur affinity.

Crocodiles, dinosaurs and early birds develop two arch-shaped openings in their skulls behind the eyes, near cheek and temple. So do lizards and snakes. Modern bird skulls with their large eye sockets and relatively large braincases tend to obliterate these openings. The openings are apparent, however, in bird ancestors. Two more openings characterized archosaur skulls: one in front of the eye socket and one in the lower jaw.

The arch-shaped openings near cheek and temple permit the attachment of jaw muscles and enhance jaw movement. The dual arches are referred to as the "diapsid" pattern—in large measure, the origin of the crocodile's bite.[5] Diapsid skulls sport two openings behind the eye socket, though the upper one is sometimes reduced during development. As the song says, never smile at a crocodile, or most of its diapsid kin. In *An Extraordinary Egg*, the chicken's mother "smiled an enormous smile," revealing its ancient diapsid heritage. The frogs failed to notice its abundance of teeth, something rare among hens.

The amnion evolved first, the diapsid skull afterwards. The diapsid pattern of skull openings preceded the origin of archosaurs, whose skulls evolved a few more openings. Archosaurs were early, but not the first, diapsids.

Diapsid structure defines a very large vertebrate group that includes, in addition to the archosaurs (for example, crocs, dinos, birds, and pterosaurs), snakes and lizards, the extinct marine reptiles (the plesiosaurs and ichthyosaurs), and the equally extinct marine lizards (the mosasaurs).

Turtles, a truly ancient lineage, are a taxonomic football, traditionally tossed into the category of anapsids, meaning "skulls without openings." The closures may be secondary characteristics, however, not the original skull condition. Recent studies based on molecular data have moved turtles into the diapsid clan and to sisterhood status with archosaurs.[6]

Think of diapsids as more closely related to one another than they are to any other amniotes. They share an ancestor in common with all amniotes, but that was a very long time ago, when tetrapod stock first crawled ashore to lay eggs in new environs. Sometime later there evolved a creature with more bite than its ancestors, thanks to those arch-shaped openings in its skull for superior muscle attachment. Its progeny (and, as now suspected, its turtle siblings) enjoyed an amniotic embryohood and diapsidly enhanced jaw function.

The evolutionary journey leading to mammals had a single opening in the skull near the eye to accommodate the jaw muscle, the "synapsid" pattern. This single opening increased in size, leaving in its wake the part of the skull known as the zygomatic arch, or cheekbone. The zygomatic arch resides just above the hinge for the jawbone. Jaw muscles pass under the arch to attach to the skull and power the snap of a dog's bite, for example. Synapsid designs and muscular jaws led to strong bites, but also to gentle suckling.

Diapsids, of course, may certainly snap their jaws tightly. Crocodilians chomp with the greatest force yet measured among living animals. The reigning bite champion, a diapsid crocodile, can clearly out-bite the synapsids' best—a spotted hyena—by a factor of 3.65. *Crocodylus porosus*, a denizen of saltwater environs, holds the current bite-strength record of 3,689 pounds of force per square inch. Of course, both groups boast ancestors with more monstrous biting abilities (*Tyrannosaurus* among the diapsids and *Smilodon*, the saber-toothed cat, among the synapsids, for instance). Among twenty-three crocodile and alligator species tested by a research team led by Gregory M. Erickson, bite force scaled to body size.[7] This correlation can be used to estimate the bite of extinct crocodilians such as *Carnufex*.

Amphibians long ago must have had to forfeit their chance to enter the bite Olympics. It's a croc specialty. Frog biting can't compete.

In summary, the amniotes are divided by a basic distinction in skull architecture. There are, of course, many other fundamental differences between synapsids and diapsids in terms of anatomy and physiology, but the mechanics of biting is a good place to start.

Hipsters and Beast Feet

Dinosaurs were a diverse lot, themselves split into many groups, though broadly divided between the "lizard-hipped" or saurischian dinosaurs and the "bird-hipped" or ornithischian dinosaurs. *Tyrannosaurus* and *Diplodocus*, for example, answered to "Hey, you, with the lizard hips!" *Iguanodon* and *Triceratops* winced when repeatedly asked, "Aren't you rather big for your bird hips?" Ironically, it was the lizard-hipped saurischians—not their bird-hipped brethren—that gave rise to the birds.

Birds were born of monsters, the theropod saurischians, to be precise. Clawed, scary-looking feet plus a furcula (primitive wishbone) make a theropod a theropod, Latin for "beast-footed." Fusion of the left and right clavicles (collarbones) forms the furcula. The modern turkey has beastly feet and a prominent wishbone, thanks to its theropod inheritance. Turkey for dinner means monster meat on the table.

Theropods assumed a two-legged birdlike stance. They were bipedal predators, big and small, built for the chase and armed for the kill.

No scene from *Jurassic Park* strikes more terror than the ground-shaking of *Tyrannosaurus rex*, a beast-footed, lizard-hipped theropod carnivore. Tiny forearms may have typified some theropods, most extremely so among the tyrannosaurs, but other groups of theropods developed longish arms for reaching and big hands for snatching.

And feathers.[8]

Richard O. Prum, curator of ornithology and vertebrate zoology at the Peabody Museum of Natural History at Yale University, and Professor of Physiology and Neurobiology Alan H. Brush have led the overthrow of the conventional interpretation of the origin of feathers. Previously, feathers were believed to be elongated, fringed, barbed reptilian scales. No longer does this hypothesis have credibility. Feathers are independently derived features orchestrated as variations of the tubal proliferation of epidermal cells. Prum and Brush explain that feathers are nourished from the inside, branched on the outside, and grow from the base, not the tip. A feather actually unfolds from a cylindrical sheath as it develops. In contrast, the top and bottom of a scale form more directly from the top and bottom of an outgrowth of the epidermis.[9]

Two genes often found guiding the development of hair, teeth, and nails also function in feather development: the gene called "sonic hedgehog" (*Shh*) promotes the proliferation of cells; bone morphogenic protein 2 (*Bmp2*) guides their differentiation.[10]

Fossil evidence complements the molecular data illustrating progressive changes in feather expression among the pre-birds. Exquisitely preserved fossils from Early Cretaceous deposits in China's Liaoning Province underscore the bird-dinosaur relationship.

Prum and Brush note that feathers, both modern and primitive, accompany a number of dinosaur fossils. A surfeit of evidence has convinced them that feathers originated among bipedal carnivorous dinosaurs prior to the evolution of birds and flight. Dinosaurs preserved in the rock strata of Liaoning range from chicken to ostrich size. Among them are *Sinosauropteryx*, a small coelurosaur discovered in 1997, which had branching tubes protruding from its skin—proto-feathers. A somewhat larger theropod, *Caudipteryx*, grew pinnate feathers on its arms and tail. Although definitely not birds, dromaeosaurs, including *Microraptor* and *Sinornithosaurus*, found at Liaoning sported feathers as well.[11]

Theropod types developed feathers long before birds existed, first at their extremities and eventually all over the body, and everywhere, to judge from their descendants as models, in gaudy colors. An assortment of sauruses, opteryxes, and raptors flashed feathery displays. Their feathered arms netted their victims and their scimitar-tipped fingers impaled them. Mix in the bloodcurdling screech of a red-tailed hawk scaled up one-hundred fold and the picture of a theropod-descended "maniraptor" (Latin for "seizes by the hand") is complete. Raptors. No wonder these killing machines, hunting in packs, became grist for the Hollywood mill. Snatching forelimbs, slashing hind limbs, mouths full of steak knives. And terribly fast.

Bone anatomy provides the clues to the close relationship between birds and dinosaurs—between extinct reptilian raptors and modern parakeets. In simplest terms, birds descended from theropods and inherited many of their traits, most directly and most recently from the maniraptors. The detailed structure of bones from the maniraptor wrist provides crucial evidence in support of this argument.

Calling a Bird a Bird

Archaeopteryx stood for some time as the earliest fossil recording a feathered bird, or at least the fossil of an extinct winged creature very closely resembling a bird.

 Technical inspection of the forelimbs of birdy *Archaeopteryx* and the maniraptor *Deinonychus* revealed a curious common trait: to wit, one semi-lunate (half-moon-shaped) wrist bone joining hands to forelimbs. By extension, the presence of this signature bone linked birds to dinosaurs. They are joined not necessarily at the hip but at the wrist. Both possessed a half-moon-shaped bone situated between the end of the arm and the beginning of the wrist.[12] It made for a very flexible joint at the base of a three-fingered hand terminating in robust claws. In fact, fossil skeletons of *Deinonychus* and *Archaeopteryx* would be difficult to distinguish without the obvious impressions of feathers in the fine-grained rock that entombed *Archaeopteryx*.[13]

In precise jargon pleasing to the ears of paleontologists and children, birds, writes Thomas R. Holtz Jr., are "coelurosaurian tetanurine theropod saurischian dinosaurs in that they have only three fingers, their tails are stiff, and the chambers in their vertebrae are very complex."[14] The vertebrate chambers are key: for birds, they provide space for air sacs that branch throughout the body (there are hollow spaces in the limbs and skull), subtract weight, aid ventilation of the lungs, and enable flight.

In order to trace the origin of avian traits, Holtz starts with the coelosaurs, the group that includes the birdy maniraptor *Deinonychus* of *Jurassic Park* raptor fame and works backwards in time. Complex vertebral chambers—air spaces in the bones of the spine—characterized the coelurosaurs; they appeared on the scene most recently in this sequence. The tetanurines stiffened their tails. They existed before coelurosaurs chambered their bones. Prior to the time of reptiles with stiff tails, the theropods stamped about on three toes. And at the beginning of this monstrous lineage stood the saurischian or lizard-hipped dinosaurs. Thus, to be quite accurate: "Birds are maniraptorian coelurosaurian tetanurine theropod saurischian dinosaurs, as revealed by their true branching feathers, their long forelimbs with half-moon shaped wrist bones [semi-lunate] in a fused hand, their big breastbones, and their big brains."[15]

That's a mouthful, but a great answer to the question "What is a bird and why do we think birds are dinosaurs?" The "maniraptoran . . . saurischian" description, punctuated by "semi-lunate" wrist bones, may invite mindless memorization. Each label, however, encapsulates a story and highlights the image of an important evolutionary innovation (the amazing feather, for example, described by Prum and Bush). The string of multisyllabic terms constitutes a chain of inferences, secured by legions of paleontologists and hordes of fossil bones, aided and abetted in interpretation by genomic wizardry. Technical jargon need not obscure the story of the descent of birds from dinosaurs. The words are a succinct and entertaining way to tell it. Erudite ten-year-olds appreciate such knowledge.

To recapitulate: birds are lizard-hipped, beast-footed dinosaurian creatures that have (or once had) paired openings for jaw attachment (the diapsid pattern), that lay calcareous eggs with amnions, that grow four limbs ("tetrapods"), that develop complexly chambered backbones, and that share a wrist bone pattern with the raptors of *Jurassic Park*. Each element of this description embodies an inference of membership within an ancestral group.[16] Backbones preceded amnions, for example, and amnions preceded diapsid skulls. The date (geologically speaking) when feathers entered this sequence remains a topic of current research, as is the anatomy of blood vessel and air space within bird and dinosaur bones, further clues to shared metabolisms.

Whatever its taxonomic accuracy, this description still fails to satisfy, as it should. It lacks what common sense and TV for children consider essential to being a bird.

Big Bird and Barney

China is the hotspot for mining fossil feathers. Feather fossils provide convincing evidence that even monstrously large "theropods"—the three-toed bipedal group that evolved tyrannosaurid dinosaurs—sported coats of "filamentous feathers" 125 million years ago.[17] These filamentous feathers may have looked very much like those on *Sesame Street*'s Big Bird. Color remains unknown, but yellow is a real possibility.

Barney, another obviously theropod-inspired creation, and Big Bird are rather closely related. "We're a happy family," Barney sings. According to anatomical criteria, birds are reasonably labeled avian dinosaurs. This categorization carries some interesting implications. For example, any regulatory policy regarding the collection of dinosaur eggs should apply to bird eggs. To date, however, farmers may gather chicken eggs without first obtaining a permit for collecting dinosaur eggs, despite the obvious ignorance of science in such policy. To common folk, if not to taxonomists, birds are birds and chicken eggs are chicken eggs. And dinosaurs are the poster children for extinction.

Unlike modern birds, *Archaeopteryx* had teeth and a bony tail, not just tail feathers. Birds grow a pygostyle, or tailbone (literally, the "rump pillar"). Modern birds retain within their DNA the instructions for growing teeth that other genes have managed to switch off. Teeth are heavy. Beaks economize weight, the better to fly. Birds today are toothless maniraptors with rump pillars.

Among modern birds, the hoatzin chicks of South America still develop clawed hands, the maniraptor pattern. Hoatzin chicks depend upon their claws to climb into the nest; these claws are lost as they develop into adults.

What, then, count as the necessary and sufficient (and not merely incidental, as feathers now seem to be) criteria for calling a bird a bird? "Avians—modern birds—have toothless beaks, enormous brains, keeled breastbones, fused wrist and palm bones, no claws on their fingers [with the hoatzin chick exception], fused hip bones, pubis [the forward bone of the hip] bones that do not meet in the middle, fused metatarsals [mid-foot bones], a first toe that points backward, and very short tails ending in a pygostyle," explains Holtz.[18] Whew.

There are two fascinating omissions in his description: *wings and feathers*. Admittedly, the deep keel on the breastbone functions to anchor powerful flight muscles, and the fused wrist and palm bones contribute to building the wing. None of these traits is found in all of the creatures nestled in the avian lineage—the 165 million years of "bird" history. Occurring together, they unite the birds of the present.

At least we can depend on birds to have an *obvious rump pill*ar. It's chic to have a pygostyle; no living bird flies around without one. The last few vertebrae fuse together to form the pygostyle and provide a place for muscles to attach that control the tail feathers. A bird's tail feathers are as essential to controlled flight as are the primaries and secondaries on the wings. The keel on the breastbone and the pygostyle at the end of the tail signify a skeleton ready for flight.

Most simply put, birds are dinosaurs with rump pillars and breastbones.

Was *Archaeopteryx* truly a bird? *Archaeopteryx* appears to have lacked the breastbone keel for attaching chest muscles powerful enough to sustain flight. Nearly two dozen vertebrae trailed out at the end of its skeleton, anchoring a frond of tail feathers rather than the modern fan.[19] Nevertheless, its feathers suited gliding flight quite well. Some birds fly; some cannot. Even Big Bird flies only on planes. Flight does not seem to be the defining characteristic of birds; *Archaeopteryx* is perhaps best classified as a taxonomic tangle.

The Dinochicken

Given that many experts consider modern birds to be avian dinosaurs, the embrace of Darwinian inquiry leads to the provocative question "Is a chicken a type of dinosaur?"[20] Maybe a chicken is more of a dinosaur than *Archaeopteryx* is a bird.

Reconstructed from imaginary fossil evidence, the "dinochicken," *Tyrannosaurus chickensis*, indicates remarkable similarities between bird and dinosaur in general appearance, despite its lack of wings. Perhaps barnyards are less domesticated than one might suppose.

Colorful archosaurs lurk among us, drawn to feeders and served at fast food chains. No wonder the frogs Jessica, Marilyn, and August mistook an alligator for a chicken. The distant ancestor of gators and birds was yet neither, but potentially both. Even the most cursory familiarity with the story of the evolution of dinosaurs leads one to watch a chicken strutting and ponder its dinosaurian affinity.

Nineteenth-century zoologists and anatomists noted obvious similarities among large birds and bipedal dinosaurs—the Big Bird–Barney correlation. For starters, their three-toed, theropodish footprints look alike. Actually, birds typically have four toes; the first toe points backwards and often is elevated off the ground.

Figure 11.1 *Tyrannosaurus chickensis*, the dinochicken. Illustration by Jan Glenn.

The three forward-pointing toes yield theropod impressions. Scales cover the legs of both groups of animals and incipient, proto-, and obvious feathers form a clear trend through time, blending bird origins with dinosaurian form. Birds lack teeth, though locked in their DNA is the recipe to grow them—or at least to grow an enamel-coated structure approximating a tooth.[21]

Any bird resembles other birds more than any bird resembles a mammal. Some birds, of course, are more closely related to each other than either is to some other bird. Is the stork more closely related to a lorikeet or to a penguin? That's a tough call. The lorikeet just seems so different from a stork . . . from beak to posture. But the penguin looks just as different—even unique among birds. Based not on skeletal similarities but on molecular data, a 2006 classification placed the stork as the penguin's closest relative, evolutionarily speaking.[22]

Genomic data from 2014 re-branched the bird tree. The newer molecular data point strongly to an evolutionary siblinghood between penguins and the tube-nosed fulmars, a marine seabird that looks like a gull. In this analysis, the next most closely related groups to penguins consist of pelicans, herons, storks, ibises, and cormorants—putting a bit of distance between penguins and storks.[23] So if the triplet is stork, penguin, and parakeet and the question is "Which one just doesn't belong?" the reasonable answer is "parakeet."

Keep in mind that the stork-penguin-fulmar-parakeet conundrum is simply fine-tuning within the archosaur lineage. Genetic analyses yield compelling evidence for the degree of shared ancestry among living species, but not certainty. The extinct ones, assuming their DNA has vanished, present the most vexing puzzles. Fossil bones, fossil teeth and beaks, fossil eggs, and fossil feather impressions must act as surrogates for genomes in order to tell the story of how *Carnufex carolinensis* and its ilk bequeathed to the world the plump, nattily tuxedoed, flightless avian dinosaur of the present. And its living cousin, the barnyard chicken.

Flights of Skepticism

Are dinosaurs not truly extinct, only evolved to become smaller and more feathery than in the past? Bones tell one story; other aspects of anatomy suggest something different, and skeptics deserve their due. Good science comes into focus in the context of debate. It may turn out that bird and dinosaur heart and lung structures differ to such an extent that the wrist connection will fail to make the case for the descent of birds and raptorish theropods from a common dinosaurian ancestor. After all, powered flight requires a very active metabolism.

Birds may trace an ancestry back to an archosaur branching at a very ancient time—a time before any modified archosaur had earned the title of dinosaur, a time when crocodiles and dinosaurs were still disentangling themselves along separate evolutionary pathways. That would make birds archosaurs then and now—diapsids as well, but no longer the descendants of theropods (the Barneysaurs). A deep archosaur origin would place chickens and alligators in much closer affinity—and more readily explain Jessica, Marilyn, and August's confusion.

Discovery in China of a Jurassic fossil bird (*Confuciusornis*) from 130 million years ago complicated the debate substantially, leading its discoverers to discount the usefulness of the half-moon wrist bone to establishing the lineage of birds with respect to dinosaurs. Specimens of this fossil bird featured the advanced trait of beaks—found in modern birds and absent from the toothed jaw of *Archaeopteryx*.

Finding a supposedly advanced trait such as a bird beak (rather than a toothed jaw) might be expected, given that *Archaeopteryx* lived 20 million years before *Confuciusornis*. *Confuciusornis*, however, had feathers and the wishbone needed for powered flight, but it retained, in contrast to *Archaeopteryx*, the skull openings reminiscent of primitive dinosaurs.[24]

Among bird genealogists, the search for origins and affinities goes on. Features essential for powered flight hold high importance, for example. When and how the bird skeleton developed bracing for powered flight remain poorly understood and may or may not be consistent with a dinosaur pedigree. And at least for now, the timing of beak origin and tooth loss poses a conundrum for avian dinosaur orthodoxy.

Bird and birdlike fossils continue to hold surprises. Specimens exhumed from Early Cretaceous rocks in northeast China's Liaoning Province sported feathered hind limbs, possibly well suited to giving a boost to gliding.[25] Stiff feathers of *Microraptor gui*, for example, spread out flat to the sides of its hind limbs. Did these "early birds" (more properly dromaeosorid non-avian dinosaurs—a family of fleet-footed and often feathered theropods) fly with four wings? Not necessarily. Plumage communicates many messages, some sexual, some territorial, some issuing a warning. *Microraptor gui's* feathers had color, even iridescence. Selection for feathery display, sexual or otherwise, might enhance survival while compromising flight aerodynamics.[26] The truth about four-winged birds remains a contentious issue.

Contemplation of Darwinian descent disquiets and discomforts many, yet exhilarates and animates others. Contentious debate is the norm, but not doubt about the reality of descent with modification. The world and all its inhabitants are mutable. Terrestrial vertebrates breathe with lungs because lobe-finned fish had lungs. Fins begat limbs. Gas-exchanging sacs for incubating embryonic young liberated eggs from watery habitats, making eggs confusable with pebbles.

For a frog, telling an archosaur egg apart from an amphibian one is as basic as knowing the difference between pebbles and gelatin. Jessica, Marilyn, and August flawlessly reasoned that a pebbly egg is a chicken egg and chicken eggs hatch chickens. Of course, the newly hatched chickengator had a curious affinity for water and a genuine talent for swimming. Instead of a pygostyle, the chickengator flexed a long tail. Scales, not feathers, covered its body. Jessica, Marilyn, and August clearly lacked a proper taxonomic understanding of what makes a bird a bird and a chicken a chicken. In their defense, substantial similarities unite chickens and alligators as archosaur progeny. They are both amniotes and they both do come from extraordinary eggs.

What Jessica, Marilyn, and August valued in their new friend was not as superficial as feathers or as flighty as wings. They resonated to its archosaur origins, its inner chickengator, its hidden diapsidness, and its exotic amniotic gestation. All that lay in their subconscious awareness of the archosaur archetype, an origin so separate from their own. What mattered most was that their new friend could teach them a few tricks about paddling and floating.

12

CORAL PIGS AND TIDE POOL SHEEP

Novel Selections of Behavior and Anatomy

> *"Of all things," grumbled Chester, "why on earth did I have to be a pig? A pig is no better off than a cabbage or a carrot, just something to eat. But before I end up as so much sausage and ham, I intend to try and amount to something."*
>
> *But what else could a pig ever be? That was Chester's main problem, and he turned this around and around in his head until one day it suddenly came to him: "I'll be a star in the circus!"*
>
> *"The nose," said Chester, "that's the thing. Surely my flat pig snout must be good for something." And he searched the pigpen for an idea. There was only a trough, a fence, and a mud puddle.*
>
> Bill Peet, *Chester the Worldly Pig*

World, watch out—here comes a pig with ambition. He has a flat snout and he's not afraid to use it. For too long had Chester rooted in the mud, his pig snout tuned to the scent of truffles. Chester had tired of a life of pig slop and wanted to become something more than sausage links.

He tried unsuccessfully to entertain people passing the barnyard. Wallowing in mud did not work. Having a flat nose proved his salvation. Everyone would notice a pig balancing on his snout. A stunning pig trick would be the ticket to circus fame. If seals could balance beach balls on their noses, a pig could balance himself. And balancing a fat pig was a more impressive feat than balancing a beach ball.

The best place to be seen was atop a fence post. Pigs climb poorly, so getting there was troublesome to Chester. With practice, he managed. At first, his nose stands lasted just seconds. At last, after days of struggle and falling over and over again into the mud puddle, Chester "mastered the impossible feat. He could stand on his nose for as long as he pleased."[1] In a short time he was discovered, and circus fame followed. Ironically, snout-standing alone did not secure his fame. Chester's back and sides were covered with dark blotches that formed a peculiar pattern. Unmistakably, he bore the map of the world on his hide. Perching on his nose displayed this remarkable trait. The novel use of his nose, the serendipitous map on his back, and his pigheaded determination to climb a fence post brought fame to Chester. Circus-goers singled him out as a favorite.

Chester's story contains elements of evolutionary drama. It tells of an already useful trait pre-adapted to a novel behavior. Snouts sniff, and pig snouts sniff with superb sensitivity. They are also tough. "Nothing wears like a pig's nose" was the motto of a long-ago marketing campaign for Finck's Red Bar overalls. Perhaps Chester's owner had a pair. In any event, Chester learned to balance on his nose for extended periods of time without wearing it out.

Balancing—along with climbing—added the second element of evolutionary drama. Chester's behavior was malleable, and he took advantage of barnyard circumstances—a fence to climb, for example. By honing his climbing and balancing skills, and exploiting his pig nose, Chester gained access to the world of the circus. He was no longer restricted to earning his living in the barnyard.

The world map on his back added a third element of evolutionary drama: a display attractive to others. Such a map ensured him more than fame, for it added to his potential for reproductive success—assuming some lady pig in his future would find the map adorning his flank as attractive as did the circus recruiters.

Bill Peet ought to write a sequel to his book that tells the story of subsequent generations of Chesters. Favored pigs live long and prosper—and they get to breed. Although there was no guarantee that Chester's children would have his attributes, the chances looked pretty good. At the least, he could teach his piglets snout-standing as an artifact of pig culture. Piglets with maps on their backs trained to do nose stands certainly mean profits for the savvy pig trader. As both circus performer and stud pig, Chester would indeed be quite the piggy bank.

Chester was certainly an entertaining pig as well as an evolutionarily instructive one. His story demonstrates how to put an old part to novel and profitable use and underscores the ways in which flexible behavior gives access to a new environment. The visual display on his hide led to his selection as a performer. By applying a bit of imaginative extrapolation, the evolutionary savvy reader can recognize that such a display resembles the adornments seized on by sexual selection. To the flashy in many species goes the mate and, ultimately, reproductive success. At least Chester's serendipitous combination of behavior and appearance saved him from the butcher, giving him a chance to grow old and perhaps enjoy piglet grandchildren. Imagine the pig talents and geographical pigskin designs that breeders might achieve after selecting from his offspring generation after generation.

Intertidal Sheep and Coral Pigs

Darwin, in addition to imagining bears skimming insects from the water as a model for whale ancestry, also tried to imagine how the elephant's ancestors may have varied from one another. Primordial elephants migrated out of Africa across the globe and adapted to climatic extremes from Russia to Canada. They had to have constitutions suited to extremely different climates and the guts to subsist on many different foods.

Even stranger than the global dispersal of elephants may seem the invasion of the sea by whales (or by something that became whales). Broadly considered, elephant ancestors include the creatures leading to the manatee or sea cow. Thus the elephant line claims kinship with sea creatures just as strongly as the society of whales claims a landlubbing ancestor. Chester's story helps to make these extraordinary transitions plausible by directing attention to the known foraging behavior of livestock in present time. In referring to the origin of the manatee, Thomas J. O'Shea wrote:

> We do not know what forces of natural selection drove an ancient mammal to exploit the niche of a large marine herbivore, yet the beginnings of such an ecological strategy can still be seen today. Domestic sheep off the islands of Scotland forage for marine algae in the intertidal zone, even swimming between patches; pigs of the Tokelau Islands in the South Pacific habitually forage along coral reefs, wading with heads submerged.[2]

For Chesterian pigs, fence-climbing and nose-standing equate with "the beginnings of such an ecological strategy." Lanolin-laced wool protects sheep from Scotland's damp chill. This same wool coat blocks hypothermia as they swim between patches of algae. Wool is sheep's blubber. Scotland's algae-munching marine sheep have not been optioned as a circus act. Nevertheless, imagine visiting a children's petting zoo where the sheep are swimming in an exhibit of tide pools.

As for Tokelau's pigs, they might give Sea World's trained sea lions a run for their money. The idea of turning pigs loose on a coral reef does seem rather abhorrent, but the point is that their diet and behavior are flexible enough to exploit an environment unlike mud and garbage. They forage with their heads down. Some must twist their snouts upwards in snorkel fashion. Those are the pigs that can keep their heads down longer and fatten up faster. Selective breeding of pigs from the most successful snorkelers fuels interesting speculation. Breeders might be able to increase snout length, leading to coral-feeding pigs with short trunks. Or maybe the nostrils will migrate . . . and future coral pigs will spout blowholes. No one can know where the extrapolation of a small advantage for pig snorkeling might evolve, but the example of flexible feeding behavior illuminates how opportunity may set a population adrift from its ancestral ways of living and provide natural selection a new palette for its compositions.

Behavioral Analogues

Chester's success depended on people's interest in his talent. The new directions in pig evolution suggested by Chester's story required the interaction of two species: Chester's appearance modified people's behavior just as people selected Chester for circus display. People began to pay to see a "worldly" pig with acrobatic talent. Previously, they'd ridden by the barnyard paying little attention to his antics.

Another example underscores the incipient ecological strategy that, with extrapolation, suggests evolutionary possibilities for two cooperating species. On the island of Chiloé, Chile, cows are put out to pasture near the sea and herded by dogs. Their foraging takes them into the intertidal zone. As the tide retreats, it leaves edible goodies for both dog and cow lining the beaches. What other nutritious resources might cow hooves and dog paws uncover? At present, manatees hold the title of "sea cows." Giving the herding and pasturing behavior on Chiloé, the island's cows and dogs are perched on the threshold of marine mammaldom, hungry to lay claim to the names "sea cow" and "sea dog."

In the *Origin of Species* Darwin reasoned that flexible behavior opened the door to new adaptations. At the same time, variation in structures provided opportunities to behave in novel ways. "It is, however, difficult to decide, and immaterial for us, whether habits generally change first and structure afterwards; or whether slight modifications of structure lead to changed habits; both probably often occurring almost simultaneously."[3] When a pig stands on its nose, its possible futures multiply.

Figure 12.1 Web-footed pigs (*Porkus duckfootus*) adapted to feeding around coral reefs. Illustration by Jan Glenn.

Darwin made observations of his own supporting this insight and shared several examples communicated to him by other naturalists. The tyrant flycatcher of South America, he reported, hunted in the manner of a kestrel, hovering over a spot, then, like a kingfisher, "standing on the margin of the water and then dashing into it."[4]

In his further explorations of behavior and adaptations, Darwin described how the English titmouse might climb like a creeper, kill prey like a shrike, or break open seeds like a nuthatch. He closed with the very provocative image of the North American black bear "swimming for hours with widely open mouth, thus catching, almost like a whale, insects in the water."[5] Berries, salmon, and bugs feed grizzlies too, the omnivore's dietary habits being an inviting open door to evolutionary possibility.

Though the gaping-mouthed, insect-straining "bear-whale" may remain a creature of Darwin's imagination, the image is not entirely fanciful, and pet owners might testify to the haunting similarity of puppy play to the filmed romping of wild bear cubs. Canids (dogs) and bears (ursids) appeared on the scene before the end of the Eocene. Bears masticate plants with molars more developed for grinding than those of dogs, yet the polar bear survives on a diet of fish and seal. Dogs, interestingly, do savor the perfumed scent of dead fish and relish the chance to roll their backs on a fish carcass.

As Darwin cautioned, asking which came first, the structure or the function—the tooth or the preferred taste—leads to no clear answer. Still, with many mouths to feed and limited food, there ensues a struggle for existence that turns small differences in foraging into major targets of selection.

The Metaphorical Origin of Natural Selection

Darwin, in order to form a new idea, depended on the application of metaphor and analogy. He drew a strong analogy between variation and selection under domestication and variation and consequent selection in the struggle for existence.

There is no agent, no volition, no will, no director carrying out the process of selection, despite the inherent implication that this is so in the word "select." "Selection" is metaphorical talk, derived by analogy with "selective breeding" of livestock and implying the personification of Nature. Natural selection is introduced metaphorically as a theory. To think literally that a personified Nature acting as a selective breeder gradually and in small increments improves species would be to misinterpret Darwin. The purpose of the metaphor is to draw attention to the existence of variation in nature analogous to the variation observed among domesticated plants and animals. In nature, as in domestication, new forms are derived from this variation. Natural causes achieve results similar to Nature acting as if she were a breeder. The metaphor is Darwin's way of making this process intelligible in terms of a phenomenon people know well, that is, selective breeding. There is no personified selector, however, working to improve life in Darwin's theory.

Darwin acknowledged this limitation and strongly criticized those making such an interpretation of his original idea. He dispelled any notion of a choice being made among animals and plants or of Nature exercising a form of will or implementing a design. Adaptations are not purposeful because the variations on which they depend come about without any correspondence to the problems of survival faced by an organism. Darwinian selection occurs after the fact of variation; it does not cause the variation upon which it acts. And yet it does not act, except in a metaphorical sense. He wrote to his critics and misinterpreters:

> Several writers have misapprehended or objected to the term natural selection. Some have even imagined that natural selection induces variability, whereas it implies only the preservation of such variations as arise and are beneficial to the being under its conditions of life. No one objects to agriculturalists speaking of the potent effects of man's selection; and in this case the individual differences given by nature, which man for some object selects, must of necessity first occur. Others have objected that the term selection implies conscious choice in the animals which become modified; and it has even been urged that, as plants have no volition, natural selection is not applicable to them! In the literal sense of the word no doubt natural selection is a false term;—but whoever objected to chemists speaking of the elective affinities of the various elements? And yet an acid cannot strictly be said to elect the base with which it in preference combines. It has been said that I speak of natural selection as an active power or Deity; but who objects to an author speaking of the attraction of gravity as ruling the movements of the planets? Everyone knows what is meant and is implied by such metaphorical expressions; and they are almost necessary for brevity. So again it is difficult to avoid personifying the word Nature; but I mean by Nature, only the aggregate action and product of many natural laws, and by laws the sequence of events as ascertained by us. With a little familiarity such superficial objections will be forgotten.[6]

Even Alfred Russel Wallace, coauthor with Darwin of the paper read on July 1, 1858, before the fellows of the Linnean Society in London, objected to the phrase "natural selection" (or, as alternatively phrased early on, "selection by natural means"). He preferred "survival of the fittest," a phrase appropriated from the English philosopher and sociologist Herbert Spencer, which Darwin came to accept as a synonym.

 Selection is passive, not an active power or force, a summation of natural laws and the sequence of physical events that conform to them. Had Darwin been more mathematically inclined, he might have fashioned his metaphor from gambling. Natural selection is shorthand for the probability of reproductive success correlated with different adaptations or traits (structures and their functions). "Favourable variations," as Darwin called them, are those that improve the probability of reproductive success for an individual. "Injurious variations" reduce this probability.

Darwin understood that variations might prove injurious or favorable depending upon context, or the "the organic and inorganic conditions of life." These conditions change on many scales in time and place, altering the probabilities that any given adaptation might prove favorable or harmful. Natural selection is a process of gambling reproductive success on patterns proven successful in the past, yet hedging the bet with heritable variations resulting from chance. Of course, there is no gambler placing the bet.

The Creative Power of Sexual Selection

Interestingly, there is one very substantial example of there indeed being a selector doing the selecting: choosing a mate. Mate selection, unless randomized, is a process based on choice, as Darwin clearly recognized in the *Origin*. Choosing a mate means sizing up the potential for reproductive success. The act of securing a procreative partner, therefore, acts as a surrogate to natural selection. Darwin labeled this process "sexual selection."[7]

Natural selections works through the struggle to secure resources needed for survival (e.g., food) in competition with others. Sexual selection operates in the context of the struggle among the individuals of one sex, whether male or female, for the chance to mate with an optimum partner. Songs, territory, or color may do the trick. Sometimes competition among males leads to an arms race that culminates in oversize tusks and antlers. The results of losing are not death for the unsuccessful competitor but few or no offspring. Sexual selection is therefore "less rigorous than natural selection."[8]

Peacock tail feathers are the most commonly cited example of male display subject to female selection. Presumably, removal of feathers and eyes from the peacock's tail would lessen peahen amorousness. Perhaps song was language's precursor among pre-humans, functioning as a lover's aphrodisiac in much the same manner that the sonorous bellowing of a swamp toad charms a potential mate. Why does a firefly flash at risk of falling prey? For love, no doubt; selection of males by females drives this risky behavior.

While Darwin posited sexual selection as a means of explaining sexual dimorphism (the disparity of appearance between males and females of the same species), evolutionary biology successors have found sexual selection a potent factor in speciation, the generation of new species. For example, as told by Ernst Mayr, there were until recently over four hundred species of cichlid fish in East Africa's Lake Victoria (introduction of non-native Nile bass reduced the count). Dramatic speciation among African cichlids occurred within just tens of thousands of years or even less. As Mayr explains, female selection of mates was the driving mechanism for branching the population into new species with specific habitat preferences.[9] The females preferred to mate with males that shared their own preference for various habitats within the lake.

Habitat preference is a behavior, and one might assume that ancestral cichlid populations varied in their tendency to favor deep or shallow water, nearshore or open-water habitats. Mate selection functioned to isolate sub-populations and opened the door to natural selection being able to modify descendant populations for optimum survival in different habitats to such a degree that they no longer closely resembled their cousins nor chose to interbreed with cichlids adapted to other habitats.[10]

When the flexibility of behavior leads organisms to exploit a new resource and physical adaptations fall into place as a result of competition for it, the appearance of purposeful adaptation is virtually inescapable. Think of the tide-pool-grazing sheep and coral-foraging pigs put to sea by enterprising farmers and extrapolate thousands of generations into the future, assuming a life-or-death struggle over these novel food sources. Now imagine the liabilities of soaking-wet wool in a cold climate or the inefficiencies of dragging a pig belly over sharp coral. Consider the ways in which terrestrially adapted feet might better maneuver on shallow sea bottoms or how some teeth and jaws might function to better effect than others.

Add in the factor that females might evolve to prefer to mate with males that appear to suffer illness or starvation the least. Those lady pigs preferring pig boys with trim waists might find themselves bearing piglets less likely than their pot-bellied cousins to suffer belly wounds when foraging among shallow-water corals.

Think of how change in diet might force accommodations in teeth and jaws, gut and enzymes. Forecast that only a minority of each generation will survive to reproduce, preserving favorable variations and subtracting others from the population. Linking mating preference, at first by chance, to traits that signal fitness then accelerates evolution. Darwinism warrants the prediction that future coral reef pigs or intertidal sheep would be markedly different from, yet recognizably similar to, those of the present. That's a very imprecise prediction. There is no way to predict the precise course of evolution and the establishment of specific adaptations, despite whatever whimsical fun there is to be had in such an attempt to extrapolate the outcomes of selection pressures whether natural or sexual. The conditions of life and the randomness of variation are just too complex and uncertain. Darwinian histories have their limits.

Selection Doubters

The followers of French naturalist Jean-Baptiste Lamarck generally agreed to natural cause as the basis of evolution, but proposed explanations understood after 1859 (the year of publication of Darwin's *Origin*) to be in conflict with natural selection. Lamarck, for example, concluded that the efforts spent by adults to obtain food and escape predators led to the acquisition of traits—to improvements within their lifetimes—that then appeared, through inheritance, in progressively more advanced forms in successive generations. To a Lamarckian, individuals did in fact adapt to the conditions of life. The engine driving this process was an innate inner drive to perfection, as opposed to natural selection. Chester the worldly pig had an inner drive, for example, to achieve fame. In this sense, his story is non-Darwinian.

For Darwinists, populations adapted as a consequence of the differential survival rates among individuals who varied from one to another. Variation happened without purpose, but with the consequence, under selection, of changing populations into new species.

Skeptics of natural selection struggle to accept the adaptive value of "intermediate forms." Darwin's critics raised what I discussed previously as "the Mivart objection": Of what use is a proto-eye, an organ of sight that cannot see? Of what value is a pre-wing, a flapless structure that cannot fly? What would a pre-elephant do with a proto-trunk? How could a whale speed though the water without a flattened fluke? Intermediate forms, argued the skeptics, must necessarily exhibit abominable adaptations. Traits, they concluded, must therefore be stable, not malleable. They tried to use variation as proof that evolution was impossible, not as the source of novel adaptations over time.

Darwinian histories, now accepted as a valid mode of explanation, did not spring fully and suddenly into being. This mode of thought itself evolved. Debates about "the transmutation of species" animated Darwin's youthful studies decades prior to his realization of the role of natural selection. While a young man of twenty-three, collecting specimens in Uruguay during the early stages of the voyage of HMS *Beagle*, for example, he entered in his journal a comment about Lamarck that mixed a bit of derision with praise. The comment indicated his awareness of Lamarckian thought with regard to the mutability of species. In a passage about the tuco-tuco, a rodent encountered in abundance near Maldonado, Darwin observed:

> The Tucutuco (*Ctenomys Brasiliensis*) is a curious small animal, which may be briefly described as a Gnawer, with the habits of a mole. . . . Considerable tracts of the country are so completely undermined by these animals, that horses in passing over, sink above their fetlocks. . . . The animal is universally

known by a very peculiar noise which it makes beneath the ground. . . . The name Tucutuco is given in imitation of the sound.[11]

The noise gave the creature, a nocturnal burrowing rodent similar in its gregarious lifestyle to the prairie dog, its name. Darwin's guide told him that, for some unknown reason, many tuco-tucos are found blind.

Darwin reflected thoughtfully on what Lamarck might have made of frequent tuco-tuco blindness. First, he credited Lamarck with "more truth than usual with him" when he speculated "on the gradually *acquired* blindness of the *Asphalax*, a Gnawer living under ground, and of the *Proteus*, a reptile living in dark caverns filled with water." The creatures named by Darwin are commonly known as the mole and the cave salamander; both have reduced vision. He continued, "No doubt Lamarck would have said that the tucutuco is now passing into the state of the Asphalax and Proteus."[12]

Losing eyesight differs from gaining eyes, of course. In this instance, young Darwin, still sympathized to a degree with a Lamarckian notion of behavior being responsible for adaptive change. In Darwinian theory behavior does not cause variation. Favorable variants happen fortuitously and randomly, blind to the future. They are induced neither by the conditions of life nor the struggle of an organism.[13]

Attention to the example of the eye repeatedly surfaces in debates about natural selection. To those who find an evolving eye an organ of suspect value while intermediate in form, its intricacies are evidence of design and a reason to reject the strictly material basis of evolution outright. To Lamarckians, both good vision and blindness are acquired characteristics. Both points of view (design and acquisition) incorporate a sense of direction that contradicts Darwin's sense of natural selection—a sense of random variation stumbling its way to improvement, accumulating successes, but without design, as opportunistic as a pig standing on its nose.

Both non-Darwinian perspectives (design and acquisition) entertain a mystical dimension—a non-physical, superordinate, and final cause. For Lamarckians, direction stems from an inner drive to perfection. From the perspective of design, divine will intervenes in the natural world to cause adaptations. They imply a cosmic consciousness and add mystery to, not subtract it from, the universe. Darwinism finds direction only in hindsight. It accounts for the diversity of life, the shared features of organisms, the geographic distribution and speciation of new forms, the record of extinction and the existence of complex adaptations from chance and necessity consistent with material cause. Darwin concluded, in effect, that the "lowest" aspects of nature—its high rates of mortality, the functioning of sexuality, the interaction of mindless form with random events—had properties sufficient to accomplish the Creation.

Whether rejecting the event of evolution itself or doubting Darwin's explanation of variation under selection, his critics held in common their unwillingness to let go of a grand endgame scheme as the reason for life's intricacy and diversity. Belief in a higher purpose, in an improving order, is comforting. There are compelling psychological reasons to cling to this belief and avoid the discomforting implications of an indifferent, unplanned, materialist history of life. After all, even Darwin found progress in the history of life.

People are immersed in their attempts to respond to one another with understanding, and in doing so they look to motivation. Human behavior makes sense when the purposes it serves are apparent, whether the aims are noble or evil; behaviors with no apparent purpose seem "crazy." Seeking purpose and meaning in life makes humans human; finding purpose in the Creation that created us seems very natural. Creation without a Creator strains credulity among Darwin's most antagonistic skeptics. They find him worse than wrong: many judge Darwinism threatening and reprehensible for banishing purpose from the natural order.

Darwin's Counterpoint

Darwin countered the critics of his concepts of incipient traits, incipient species, and the power of natural selection to manufacture novelty by focusing on (1) examples of fortuitous if not random variation (the role of chance), (2) analogies between the structures observed among living species and the potential for variation among ancestral ones, and (3) the way that scaling a trait larger or smaller without substantially changing the trait itself might produce dramatically changed organisms. In effect, he argued the evolutionary potential of variation from multiple perspectives and shared abundant proof of its reality. Barnacles varied; climbing vines varied; orchids, worms, pigeons varied. No matter the group he studied, its variability was as stunning as any defining trait. In fact, defining traits themselves were statistical properties, probabilities that a member of a group would exhibit a trait. Variability made taxonomic boundaries seem uncertain and even arbitrary, if not capricious, among taxon after taxon that Darwin scrutinized, virtually obliterating the very idea of there being such a thing as a fixed type, variety, subspecies, or even species.

For Darwinists, life's forms are not the incarnations of transcendent archetypes; life follows no fixed plan nor admits to any fixed types. It is ever malleable and evolving, transitioning or ready to transition among forms, primitive ones judged so in retrospect if other traits have been derived from them.

Darwin provided his critics with plausible stories of how past variations, extrapolated to the present, could account for the transmutation of species and explain their geographic distribution. The abundance of sloths in South America living amidst a landscape replete with the fossil skeletons of several extinct gigantic species was a perfect example. Shifts in the conditions of life, he concluded, drove the mega-sloths to extinction and left the meek, cuddly sized varieties to inherit the continent.

His work with barnacles underscored the transformation of an existing structure into something novel and the reduction of one sex (the male) into something nearly unrecognizable. During the larval stage, he observed, part of the female barnacle's reproductive structure, the ovarian tube, becomes a gland that produces cement. This cement—secreted from ducts that open at the end of the antennae—glues the barnacle to a substrate. Its head stuck for the rest of life to a rock, a barnacle no longer has any opportunity to paddle or crawl with its feet.

Modified, the feet take on a new function. Exaggeration of the fine branching forms on multiple feet turns them into feathery feeding nets. Well-mannered barnacles eat with their feet. Detailed analysis of barnacles taught Darwin about the extraordinary novelty the natural modification of structures could achieve. He demonstrated how, through scaling or differential rates of growth, a trait (barnacle glands, barnacle feet) might vary and become an altogether different adaptation from what it was at another scale—or in another time. His barnacle researches put him on his way to demonstrating how "an organ could change function as an animal exploited new conditions."[14] Among mammals under selective pressure, a modified claw could become the primate's nail or a running beast's hoof. Among whales, tiny laminar structures of the mouth might transform into curtains of baleen. Even a barnyard pig might put its nose to novel use.

New structures and novel functions emerge from older ones, constrained by existing designs and their interdependency. This is the recipe for making eye, ear, nose, and throat. It is the alchemy for changing legs to flippers, arms to wings, reproductive organs to glue sticks, and gums with keratin into mouths with baleen.

At the dawn of the science of comparative anatomy, and decades before Darwin set sail on the *Beagle*, observing guts and innards—their similarities and differences—had given birth to radical ideas about the order of nature. These ideas led to discoveries about unity amidst diversity and to distinctive groupings of life deemed natural, placing, as did Cuvier, the whales among the lactating mammals rather than among the fishes. Groupings nested obviously similar creatures into larger categories similar in more general ways. The lynx (or bobcat) belonged to the family of cats, the order of carnivores, the class of vertebrates, and the kingdom of animals, for example, in the Linnaean system of living things. All creatures nested in taxonomic categories like so many Russian matryoshka dolls. Science purported to have succeeded in its quest to "carve nature at the joints," and mindful probing of nature would never be the same.[15]

Darwin's ideas of variation, descent, and selection provided the compelling explanation for why this carving worked so well.

Darwin, in discounting a higher power's mysterious role in creating living forms, contributed to making science one of society's highest forms of authority. He demonstrated how to reason convincingly about the history of life in the materialist terms of incipient organs, transitional forms, similar structures, variability, differential survival, and both natural and sexual selection—the concepts still used to construct "Darwinian histories."[16]

It helps to put Darwin's thoughts into focus with the story of an ambitious pig. Chester's tale offers an important insight into how new uses for old parts, flexible behavior, and selection act in concert. Chester's pigskin varied in a novel way. His pattern of coloration corresponded to a map of the world. Chester learned to stand on his nose and catch the attention of circus performers. Nose-standing and pigskin map, one trait learned and the other inherited, opened a new niche to him where he prospered. Variation carried him from barnyard to circus.

In the real world there are circus pigs, of course. There are also cows that comb beaches, sheep that graze in tide pools, and pigs that root in coral. Their behavior is more than just quirky. It enables them to take advantage of novel diets in unusual habitats. Competition in these new environs changes the cluster of traits deemed advantageous or injurious, providing grist for natural selection's mill. Smart mating moves the process along. Who knows where these novel behaviors may lead in the circus of life?

EPILOGUE

FEMURS AND FOOTPRINTS

On the Trail of Megabeasts

Beneath the May Colorado sky on the bank of the Purgatoire River not far from Bent's Old Fort, a dozen third- and fourth-graders and I gathered for lunch. Seated in fossilized dinosaur tracks, we let the gray-green mudstone serve as our table. Nestled amidst the impressive footprints, the children turned their attention from eating a meal to imagining becoming one. A small herd of elephantine Jurassic sauropods had trod this way. Might a fearsome *Allosaurus* lurk round the bend, stalking them from behind?

Fossil footprints leave indelible impressions, and fossil skeletons amplify this wonder, whispering memories of life long past: of the first creatures to walk on land, of whales strolling to the sea, of the first birds, of the origins of people. Winding along erosion-exposed rock, the five sets of Purgatoire tracks extended upstream and downstream from the picnic site, encoding patterns of behavior on a single steamy Late Jurassic day. These beasts, of course, walked an ancient lakeshore long departed,[1] but the presence of the tracks was so arresting that, for a moment, the slurping and slushing sounds of dinosaurs plodding through the mud seemed to fill the air.

"I've Got an *Allosaurus* in My Bedroom, Do You Want to See It?"

On that spring afternoon, I could not have guessed that one of my students would later achieve a modicum of paleontological fame. She was a connoisseur of the Morrison Formation, able to spot fossilized bone fragments with ease.

India had been excavating fossils on successive summer visits to a friend's ranch since she was twelve. After six summers, eighteen complete *Allosaurus* bones, treasures liberated from Jurassic shale, had taken possession of her bedroom. Years later she described the scene: "Dinosaur fossils were strewn all over the place, on bookshelves, across my huge desk, and under my bed. . . . They would just fit into a bathtub."[2]

The bedroom was turning into a geologic formation in its own right. There was no more room under the bed or anywhere else for more bones.

Figure E.1 *Brontopodus* trackways along the Purgatoire River. Photograph by the author.

Figure E.2 Bones of the fearsome *Allosaurus* lurk beneath a child's bed. Illustration by Jan Glenn.

Now that India was preparing to leave for college, her mother, while impressed with her daughter's discoveries, was unhappy with the mess of chiseled-off rock, femur fragments, oversized vertebrae, and spilled glue littering the carpet. The *Allosaurus*, a once terrorizing monster that for many years had been India's childhood friend, had to go.

India appealed to the Denver Museum of Nature and Science:

> I put three bones in a shoe box, along with a note requesting they please confirm my *Allosaurus* identification, and left the box up at the museum on a Friday afternoon. The staff paleontologist called Monday and said my identification was correct, though he had at first thought my note was a joke and expected cow bones when he opened the box.[3]

Duly impressed by the extinct foundling left at the museum, the staff paleontologist came to her home to verify her claim. He entered a room full of plastic horses where India shared "box after box of beautifully collected dinosaur bones from beneath her bed."[4] He did not notice a display of fifty different brands of empty beer bottles or her high school rodeo jacket on the wall. India recalls: "I then took the paleontologist up to the site, where Don Lindsey [of the Denver Museum] told me they really wanted the specimen. I said if they wanted my

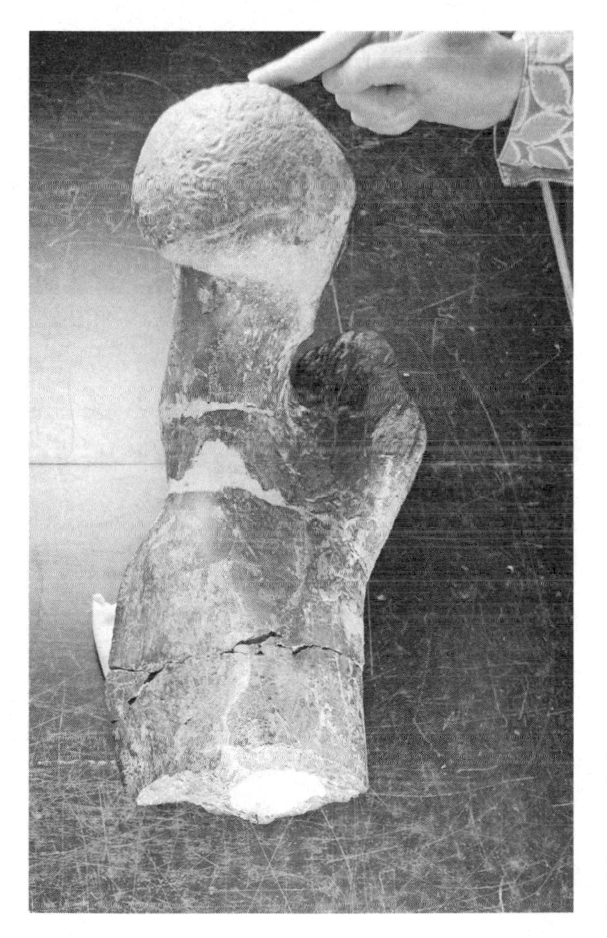

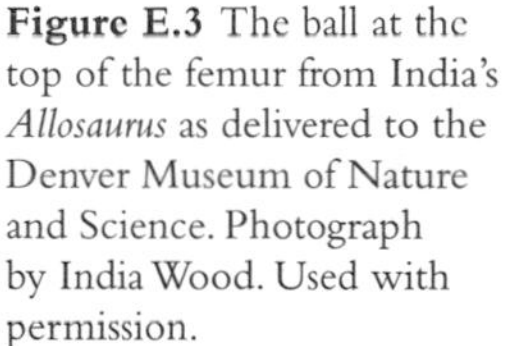

Figure E.3 The ball at the top of the femur from India's *Allosaurus* as delivered to the Denver Museum of Nature and Science. Photograph by India Wood. Used with permission.

bones, they needed to hire me as a field assistant. After apparently much discussion about whether a girl could be an adequate field assistant, they hired me."[5]

India proved to be an excellent field assistant. Soon the museum was displaying her remarkable collection as part of its "Prehistoric Journey" exhibit. Jurassic bones, lovingly collected and once stashed beneath a child's bed, now delight thousands of museum visitors daily. India's *Allosaurus* chases *Stegosaurus* in perpetuity, and parents can purchase plastic *Allosaurus* claws at the museum gift shop to decorate their own children's bedrooms.

India took after Charles Darwin. She was an inveterate collector of fossils, shells, animal bones, dead insects, and arrowheads. At seven she enjoyed gluing broken pieces of cow bones back together for classroom display and mounting grasshopper body parts in her nature journal.

The young Darwin may not have stashed dinosaur fossils in his home, but he was no slouch as a collector of beetles, barnacles, and birds, not to mention the bones of "antediluvian" beasts. India must have wondered just what precious mementos Darwin kept beneath his hammock on the *Beagle*. "I've stuffed the *Beagle* with the claws of giant ground sloths and the carapaces of monstrous tortoises," he might well have written to his mentors at Cambridge University and the British Museum. "Do you want to see them?"

Patagonian Megabeasts

Darwin reveled in the extinct megabeasts of South America, creatures that burst open his imagination. From the cliffs at Punta Alta, the riverbanks of the Argentine pampas, and the terraces of Patagonia, Darwin dug from the ground fierce skulls, enormous claws, and stupendous teeth. The soft sediments of the pampas entombed the tank-sized bodies of *Hoplophorus* and *Glyptodon*, cousins of armadillos decked out in polygonal-plate armor.[6] Defensive measures on this scale meant that monstrous predators—some really nasty saber-tooths?—once lurked about.

Among his most remarkable finds towered four giant ground sloths: *Scelidotherium*, *Mylodon*, *Glossotherium*, and *Megalonyx*. Of the diverse assemblage of species related to the giant ground sloths (the "xenarthran" mammals, powerful diggers with strongly reinforced lower backbones), only much smaller tree sloths, anteaters, and armadillos exist today.

Megatherium, with its seven-inch claws, was the scariest ground sloth of all, and the preposterous *Toxodon* mixed rhino legs with a hippo's maw. *Toxodon* looked something like a flat-footed manatee with rat's teeth crossed with a hippo-headed rhinoceros. In other words, just plain weird. Its nostrils and ears suggested to Darwin that *Toxodon* lived a semi-aquatic lifestyle, lumbering around like a "rhinoppotaratamatee." Darwin extolled the significance of this hodgepodge of a beast, writing, "How wonderfully are the different Orders, at present time so well separated, blended together in different points of structure of the Toxodon!"[7]

Several times he made taxonomic errors that England's anatomists had no trouble correcting (FitzRoy recruited him as a geologist, not a biologist). At one point Darwin speculated on the significance of fossil bones he had unearthed to the north of Punta Alta. He interpreted them as the remains of a long-necked, three-toed, horse-sized beast—*Macrauchenia*. Perhaps humpless, or perhaps amply humped (no one knows for sure from the fossils), *Macrauchenia* with its long neck resembled a camel; oddly, the creature had a skull that suggests the presence of a tapir-like elongate flexible snout. Maybe it deserved the name "camorse," of course. Darwin thought he had discovered the ancestor of the living guanaco (a llama), a member of the very dignified camel family. Camels, however, are quite distinct from macrauchenids, though not nearly as different as rodents are from elephants.

Macrauchenia was indeed strange-looking. Think of an otherworldly steed in a *Star Wars* movie. This beast was the last survivor of a group of mammals restricted to South America: the litopterns (semi-hoofed mammals with simple ankle bones).

On another occasion, Darwin thought he had found the skull of a rather large rhinoceros. Alas, more expert analysis revealed the skull to be from a *Scelidotherium*, a three-ton Patagonian ground sloth.

Yet Darwin's sloth-sense was improving. From tooth and jawbone he correctly identified a *Megatherium*. "I obtained a jaw bone which contained a tooth," he recorded in his *Beagle* diary. "By this I found out that it belongs to the great ante-diluvial animal the *Megatherium*."[8] Patagonia harbored the remains of this giant ground sloth almost in abundance.

Figure E.4 An artist's rendering of *Macrauchenia*. Perhaps it had a hump on its back like a camel. Illustration by Jan Glenn.

Figure E.5 *Megatherium* imitating a small child. Illustration by Jan Glenn.

In all, he returned to England fossils of "nine great quadrupeds" with resemblances to horses, camels, pachyderms, rhinos, armadillos, and giant rodents (the "Gnawers"). The strangest among them was *Toxodon*, perhaps as closely related to *Macrauchenia* as rhinos are to horses. Despite his inferential errors, he achieved stunning insights. The fossil record of faunal successions and faunal distributions—the wonderful relationship between dead and living—has indeed thrown "light on the appearance of organic beings on our earth."[9]

Fossil "Ostrich" Tracks

In 2003 I journeyed to Argentina to join my son on the trail of Darwin's megabeasts. We learned that populating the ancient landscapes of South and Central America were a very diverse group of semi-hoofed creatures including the bizarre and hefty-toothed *Toxodon*, plus *Macrauchenia*, the camelish horse of fleshy proboscis fame.

Each of these extinct beasts carefully described by Darwin left three-toed tracks on the elevated Patagonian terraces. Eager to replicate a small portion of his discoveries, we avidly hoped to find their footprints, if not their bones.[10]

We drove to Pehuen-Có and arrived to find ourselves the sole guests at the beachfront lodging, Hotel Cumelcan. Cold, moist air blew in from the Atlantic. It was late March, with fall steadily creeping into the southern latitudes. The hotel owners lit a fire in the dark, empty dining hall. We huddled in wooden chairs around the warmth and light.

The next day, wrapped in the chilly fog of early morning, we piled into a jeep to head south with our guide at the wheel. The vehicle crawled along the wet beach sand over pieces of polygon-cracked fossil mud and faded Pleistocene oyster beds. Once disembarked, we examined a few exposed surfaces of prehistoric mud protruding above the sand, finding only the tracks of the still extant flamingo and guanaco. In short order we imagined that indistinct yet obvious depressions in well-weathered and supposedly older mudstone were the sought-after apparitions: imprints of multiple megatheria crossing a vast Pleistocene savanna.

Alas, our guide informed us that they were just weathered puddles. "If we dug up two feet of sand," the guide explained, "here is where the footprints would be." We took little satisfaction as our tour ended without our having walked in any footprints of *Megatherium*.

We also missed the evidence in the mudstones from a few thousand years ago of children absconding with flamingo eggs. Recent investigators have uncovered fossilized human footprints at Monte Hermoso. During the Pleistocene, as sea level dropped and shorelines receded, shallow lakes and marshes dotted the pampean landscape where flamingos, guanacos, and even megatheria once roamed.[11] Imagine children out foraging for flamingo eggs, wary of meeting a giant ground sloth with seven-inch claws.

Coffee and croissants at the hotel enlivened our spirits as we headed next to the Museo Carlos Darwin. Just blocks from Punta Alta, the museum explains a great deal of pampean fossil history to the public and visiting schoolchildren alike. The museum has published a field guide to the fossil beds of Pehuen-Có and Monte Hermoso.[12]

A series of exhibits teaches that by 3 million years ago, an island chain had solidified into the present-day land bridge between the Americas. Sloths and armadillos headed north, while woolly mammoths, saber-toothed tigers, ancient horses, and cameloids (llamas, for example) reached South America. During the Pliocene, the earth cooled, ice sheets formed, and sea levels dropped. The Pleistocene saw global sea levels rise and fall with recurrent glaciations, the most recent of which left the horizons of fossilized, cracked mud—the formations we saw at Pehuen-Có. When here, Darwin exclaimed, "What a history of geological changes does the simply-constructed coast of Patagonia reveal!"[13]

As tall as an African bull elephant rearing up on its hind legs, a life-sized replica of *Megatherium* keeps watch over the main gallery at Museo Carlos Darwin, where children can pose in terror while pretending to gather flamingo eggs. It is, as Darwin claims, impossible not to feel deep astonishment for a time when these monsters lumbered across the Pampas alongside battle-tank armadillos (admittedly, small tank size) and the mysterious *Toxodon*, "one of the strangest animals ever discovered."[14]

Figure E.6 A young boy unintimidated by the skeleton of *Toxodon*. Illustration by Jan Glenn.

Armed with the prior knowledge afforded by our visit to the museum, we worked our way south to Carmen de Patagones and the Rio Negro, then accepted an invitation to stay in Viedma at the home of a motorcycle enthusiast on the south bank of the river. Our host mentioned some interesting fossil-laden cliffs at a place called Playa Bonita, which became the next stop on our megabeast quest.

Playa Bonita's eighty-foot cliffs towered above the murky ocean waters. At their base we encountered a fossilized Miocene oyster bed with shells caked in green clay. Just above this layer, remnant horizons of an ancient savanna disappeared below the sandstone. Were it not for the presence of fossils, these layers might have appeared to have dried up within the last few months.

Here our zeal began to skew our observations. Among the terraced cliffs we found a plethora of footprints and puzzled over their origin. A few we confidently identified as two-toed ungulates representative of the modern order Artiodactyla—deer, camels, sheep, pigs, cattle. Most likely we had witnessed the passing of a guanaco (wild llama). Nearby lay the tracks of a few flamingos.

Elongate, three-toed, deep-set, moderately large impressions captured our attention. These, we surmised with some joy, were the traces of some wandering *Macrauchenia* or possibly a *Toxodon*. At least we wished to believe so. The tracks of an extinct Patagonian beast! We knew the strata came from the right horizon of time, that Darwin had found fossils of these creatures amidst other eroded Patagonian cliffs, and that three-toed prints were emblematic of these enigmatic creatures.

Five years later we looked again at our photographs. Doubt entered our minds. Had we inferred poorly and with more assurance than our knowledge warranted? Like Darwin, we began to suspect "a good deal of the Dousterswivel" in ourselves.[15]

Something about the tracks had always puzzled us. In a way, they seemed vaguely familiar and not quite so alien as a splayed-toed quasi-ungulate. What was it they reminded us of? . . . India's *Allosaurus*! They looked like the tracks of a dinosaur. And for good reason.

Figure E.7 Theropodish *Rhea americana* fossil footprint at Playa Bonita, Rio Negro, Argentina. Photograph by the author.

Googling fossil dinosaur footprints taught us that the modern rhea, the ostrich-like bird from Patagonia, possesses a hind limb structure similar to that of the bidpedal theropod dinosaurs. It leaves three-toed tracks. In other words, the modern rhea provides a living analogue for how *Allosaurus* likely ran. The limbs, feet, and gait of rheas (and emus) apparently work well as analogues to the primordial stock that bequeathed birds to the world. And *of course* their footprints resemble each other.

Were the tracks, in fact, not from the extinct monster of our (and Darwin's) imagination, but instead the footprints of the familiar large ostrich-like flightless bird, the rhea? We had seen many of them near Torres del Paine. Searching online for photos of rhea footprints confirmed our hunch. Indeed, these were rhea tracks in Pleistocene sediments. Unmistakably, we had been on the trail of a rhea—perhaps *Rhea darwinii*, the bird virtually unknown to Western naturalists in the 1830s which Darwin inadvertently ate.

Although Dousterswiveled while in Patagonia, we found ourselves excited to be able to distinguish the tracks of three-toed bipedal birds from those of three-toed quadruped ungulates—a rather amateur accomplishment, to be sure, but deeply satisfying nonetheless. The Pleistocene's *Rhea* ran with the *Macrauchenia* and *Toxodon* crowd, and we now, with some assurance, knew the difference. Our visit to the garden of Darwinian delights had reminded us of the wonderful relationship between dead, terrible, giant lizards and living, flightless birds.

Revering the Living

Living jaguars, pumas, guanacos, rheas, and rodents of unusual size greeted Darwin in Patagonia. Darwin, as papa and grandfather, naturally told stories to his children and grandchildren that underscored the astonishing diversity of life. Perhaps they stashed sharp teeth, terrible claws, and strange bones among their playthings.

Reconstructed by artists, *Toxodon, Macrauchenia, Glyptodon, Megatherium*, and *Scelidotherium* roam today's museum galleries, appearing as so many monsters from the id. Playfulness with megabeast imagery is indeed serious fun, and the names of South America's megabeasts are silly enough to enjoy saying out loud again and again. *Toxodon, Toxodon*, oh mighty *Toxodon*, thou noble notoungulate! Similarities and differences among ankles and elbows, wrists and toes—and even footprints—have tales to tell about the connections between living things and their ancestors. Whimsical takes on stories of origins may please preschooler or postdoc while at the same time inviting a careful look at evolution.

Decades ago a young child visiting the Denver Museum of Nature and Science with her mother entered the hall dominated by the skeletons of dinosaurs. The child shivered, then nearly froze in terror. Calmly, the mother tried to explain extinction and counseled the little girl not to worry. "All the dinosaurs are gone now," she said. The child clutched her mother's hand ever more tightly and faintly asked, "But Mommy, why did they all die *here*?"[16]

They didn't of course and given the child's lack of knowledge her fright was quite understandable. Some years later, an adolescent, who had picnicked in the footsteps of dinosaurs as a fourth-grader, diligently collected fossil bones as a pastime over several summers. Now mounted at the Denver museum, India's *Allosaurus's* scimitar claws and scissoring teeth threaten the life of its startled prey, *Stegasaurus*.

Darwin's science, its imagery and stories, enlightened and enlivened her life. As a parent, she shared the landscape and excitement of her youthful collecting adventures with her own children, as did Darwin, expecting them to acquire some of the same joy she found in the natural world.

Of course, not everyone has the privilege of sleeping with an *Allosaurus* skeleton stowed beneath the bed. Yet, with guidance, a child's fascination with life's extinct megabeasts and Jurassic fauna may reinforce appreciation, even reverence, for living creatures great and small. That's the most important lesson to learn from Darwinian stories about the "wonderful relationship . . . between the dead and the living."[17]

NOTES

Introduction

1. Stephen Klassen, "The Relation of Story Structure to a Model of Conceptual Change in Science Learning," *Science & Education* 19 (2010): 305–17. Klassen analyzes the importance of story form to explanation with an emphasis on story being a means to re-create or get close to the events of interest.

2. Jacob Bronowski, *The Ascent of Man* (Boston: Little, Brown, 1973), 153.

3. Rudyard Kipling, *Just So Stories* (1902; Garden City, NY: Doubleday, 2002), 63.

4. Robert Louis Stevenson, *Treasure Island* (1883; Santa Rosa, CA: Classic Press, 1968).

5. Flustrids belong to the Bryozoa, a name that means "moss animals."

6. Charles Darwin, *On the Origin of Species by Means of Natural Selection, Or the Preservation of Favoured Races in the Struggle for Life*, 1st ed. (London: Murray, 1859), 484, in *The Complete Work of Charles Darwin Online*, ed. John van Wyhe (2002), http://darwin-online.org.uk/.

7. *The Autobiography of Charles Darwin*, ed. Nora Barlow (New York: W. W. Norton, 1969), 119.

8. Ibid., 118–19.

9. Karen Gallas, *Talking Their Way into Science: Hearing Children's Questions and Theories, Responding with Curricula* (New York: Teachers College Press, 2002), 26.

10. Ibid., 41–47.

11. Joseph J. Schwab, "The Concept of the Structure of a Discipline," *Educational Record* 43 (1962): 198.

12. Klassen, "Relation of Story Structure to a Model of Conceptual Change," 312.

13. "Tetrapod" refers to vertebrate animals with paired limbs and literally means "four-footed." It can also refer to the descendants of the first tetrapods that have subsequently evolved limbless bodies (snakes) or lost their hindlimbs (whales). The first back-boned forms of life to crawl on the land earned the moniker "tetrapods."

14. Leo Lionni, *Fish Is Fish* (New York: Pantheon Books, 1970).

15. Bernard Wiseman, *Morris the Moose* (New York: Harper & Row, 1989).

16. Graham D. Burnett, *Trying Leviathan: The Nineteenth-Century Court Case That Put the Whale on Trial and Challenged the Order of Nature* (Princeton: Princeton University Press, 2007), 4. Burnett drew heavily upon the historical *Maurice v. Judd* trial as recorded in William Sampson, *Is a Whale a Fish? An Accurate Report of the Case of James Maurice against Samuel Judd* (New York: Van Winkle, 1818).

17. Phillip Kitcher, *The Advancement of Science: Science without Legend, Objectivity without Illusions* (New York: Oxford University Press, 1993), 27.

18. Charles Darwin, *The Origin of Species by Means of Natural Selection, Or the Preservation of Favoured Races in the Struggle for Life*, 6th ed. (London: Murray, 1872), 402, in van Wyhe, *The Complete Work of Darwin Online*.

1. "Curtiosity's" Child

Elements of this chapter first appeared in Charles R. Ault Jr., "On the Origins of Darwin's Impertinence," *American Paleontologist* 18, no. 1 (2010): 29–31. Reprinted with permission from the Paleontological Research Institution, Ithaca, NY.

1. Ernst Mayr, *What Evolution Is: From Theory to Fact* (New York: Basic Books, 2001), 11.

2. Ibid. Note that these words refer to Darwin's thinking as quoted in the epigraph to the introduction of this book. Darwin surmised that similarities between creatures in the fossil record and the living biota of the same region would provide fruitful data bearing on the question of origins. His principle is a sound starting point, but not a definitive one. Animals may supplant others by migration, for example. Because of his limited perspective, Darwin mistakenly inferred the extinct *Macrauchenia* to be a relative of the modern llama and believed that *Toxodon* held an analogous relationship to capybaras. At the same time, this geographic style of hypothesizing helped to bring evolutionary ideas into focus.

3. Adrian Desmond and James Moore, *Darwin: The Life of a Tormented Evolutionist* (New York: W. W. Norton, 1991), 13.

4. Ibid., 14, 16.

5. Janet Browne, *Charles Darwin*, vol. 1, *Voyaging: A Biography* (Princeton: Princeton University Press, 1995), 29–34.

6. Desmond and Moore, *Darwin*, 7.

7. Ibid., 29, 229.

8. Randal Keynes, *Darwin, His Daughter, and Human Evolution* (New York: Riverhead Books, 2001), 108.

9. Ibid., 110.

10. Jean-Jacques Rousseau, *Émile or On Education* (1762), trans. Allan Bloom (New York: Basic Books, 1979), 37.

11. Frank Doherty, "The Wedgwood System of Education," *The Wedgwoodian* (November 1983): 182–87, quoted in Keynes, *Darwin, His Daughter, and Human Evolution*, 111.

12. Johann H. Pestalozzi, *Wie Gertrud Ihre Kinder Lehrt* [How Gertrude Teaches Her Children] (Bern: Gessner, 1801).

13. Keynes, *Darwin, His Daughter, and Human Evolution*, 111.

14. Ibid., quoting Pestalozzi from E. Woodall, "Charles Darwin," *Transactions of the Shropshire Archaeological and Natural History Society* 8, (1884): 14.

15. Keynes, *Darwin, His Daughter, and Human Evolution*, 112.

16. Irving Stone, *The Origin: A Biographic Novel of Charles Darwin* (New York: Plume Book, 1980), 499.

17. Ibid., 498.

18. Jean Paul Friedrich Richter, *Levana, or The Doctrine of Education* (Boston: D. C. Heath & Co., 1890), 152.

19. Desmond and Moore, *Darwin*, 34.

20. *Charles Darwin's Notebooks from the Voyage of the* Beagle, ed. Gordon Chancellor and John van Wyhe (New York: Cambridge University Press, 2009), 38.

21. Jared Diamond, foreword to Ernst Mayr, *What Evolution Is*, vii.

22. Quoted in Michael Powell, "A Knack for Bashing Orthodoxy," *New York Times*, September 20, 2011.

23. Jacob Bronowski, *The Ascent of Man* (Boston: Little, Brown, 1973), 153.

24. Richard Dawkins, *The Ancestor's Tale: A Pilgrimage to the Dawn of Evolution* (New York: Houghton Mifflin, 2004), 362.

2. Darwin and the Pampas Pirates

Elements of this chapter first appeared in Toby R. Ault and Charles R. Ault Jr., "On the Trail of Darwin's Megabeasts," *American Paleontologist* 17, no. 1 (2009): 16–19. Reprinted with permission from the Paleontological Research Institution, Ithaca, NY.

1. Robert Louis Stevenson, *Treasure Island* (Santa Rosa, CA: Classic Press, 1968), 5.
2. *The Goonies* (1985), directed by Richard Donner (Burbank, CA: Warner Bros. Entertainment, 2007), DVD.
3. *Pirates of the Caribbean: The Curse of the Black Pearl* (2003), directed by Gore Verbinski (Burbank, CA: Walt Disney Pictures, 2003), DVD.
4. Peter Nichols, *Evolution's Captain: The Story of the Kidnapping That Led to Charles Darwin's Voyage Aboard the Beagle* (New York: Perennial, 2004), 50.
5. Ibid., 52.
6. Ibid., 52–60.
7. Ibid., 68.
8. Robert FitzRoy, Narrative of the Surveying Voyages of His Majesty's Ships Adventure *and* Beagle *Between the Years 1826 and 1836, Describing Their Examination of the Southern Shores of South America, and* The Beagle's *Circumnavigation of the Globe. Proceedings of the Second Expedition, 1831–36, under the command of Captain Robert Fitz-Roy, R.N.* (London: Henry Colburn, 1839), 4–6, in The Complete Work of Charles Darwin Online, ed. John van Wyhe (2002), http://darwin-online.org.uk/.
9. Adrian Desmond and James Moore, *Darwin: The Life of a Tormented Evolutionist* (New York: W. W. Norton, 1991), 46.
10. *Charles Darwin's* Beagle *Diary*, ed. Richard D. Keynes (Cambridge: Cambridge University Press, 2001), February, 28–29, 1832, 42.
11. Ibid., March 4, 1832, 43.
12. *Charles Darwin's Zoology Notes and Specimen Lists from H.M.S.* Beagle, ed. Richard D. Keynes (Cambridge: Cambridge University Press, 2000), 28.
13. John Kricher, *A Neotropical Companion: An Introduction to the Animals, Plants, and Ecosystems of the New World Tropics* (Princeton: Princeton University Press, 1997), 133.
14. *Darwin's Zoology Notes*, 25–26.
15. Ibid., 34.
16. Richard Keynes, *Fossils, Finches, and Fuegians: Darwin's Adventures and Discoveries on the* Beagle (New York: Oxford University Press, 2003), 76. Keynes cites Philip G. King's 1890 account of the voyage, "Reminiscences of Mr Darwin during the voyage of the Beagle," in van Wyhe, The Complete Work of Darwin Online.
17. Stevenson, *Treasure Island*, 139–40.
18. Desmond and Moore, *Darwin*, 112. The postillion was the rider of the leading horse in a coach pulling team; English gossip about the servant class often alluded to liaisons between the housemaid and the postillion.
19. Ibid., 121.
20. *Charles Darwin and the Voyage of the* Beagle, ed. Nora Barlow (London: Pilot Press, 1945), 62, in van Wyhe, *The Complete Work of Darwin Online*.
21. Beagle *Diary*, May 25, 1832, 67.
22. Ibid., April 8, 1832, 52.
23. Ibid., April 12, 1832, 56.
24. Ibid., April 8, 1832, 53.
25. Ibid., January 30, 1832, 31.
26. *Darwin's Zoology Notes*, 107.

27. Beagle *Diary*, July 22, 1832, 83.
28. Ibid., July 26, 1932, 85.
29. Ibid., August 2, 1832, 87–89.
30. Keynes, *Fossils, Finches, and Fuegians*, 93.
31. Beagle *Diary*, August 5, 1832, 89.
32. Stevenson, *Treasure Island*, 120–21.
33. Beagle *Diary*, August 6, 1832, 90.
34. Ibid.
35. Ibid., 93.
36. Ibid., August 11, 1832, 92.
37. Ibid., November 6, 1832, 115.
38. Ibid., November 9, 1832, 116.
39. Keynes, *Fossils, Finches, and Fuegians*, 103.
40. Beagle *Diary*, September 7, 1832, 99–100.
41. Ibid., 101. Keynes cites the phrase "un naturalista" from Robert FitzRoy, *Narrative of the Surveying Voyages of His Majesty's Ships* Adventure *and* Beagle, 103–4.
42. Beagle *Diary*, September 7, 1832, 100–101.
43. Ibid., May 9, 1833, 154.
44. Pharmaceutical Society of Great Britain, "On the Composition of Instantaneous Light Matches," *Pharmaceutical Journal: A Weekly Record of Pharmacy and Allied Sciences* 12 (1853): 426–28.
45. Beagle *Diary*, May 9, 1833, 155.
46. Ibid., May 10, 1833, 156.
47. Ibid., July 2, 1833, 161.
48. Ibid., August 19, 1833, 174–75.
49. Ibid., August 23, 1833, 177.
50. Ibid., September 4–7, 1833, 180–81.
51. Charles Darwin, *The Voyage of the* Beagle*: Darwin's Five-Year Circumnavigation* (1845; Santa Barbara: Narrative Press, 2001), 74.
52. Beagle *Diary*, September 4–7, 1833, 179.
53. Ibid., August 16, 1833, 172.
54. Ibid., September 18, 1833, 192.
55. Keynes, *Fossils, Finches, and Fuegians*, 199. Horses, seven stallions and five mares, were first brought to South America in 1535, to Buenos Aires.
56. Beagle *Diary*, October 21, 1833, 197.
57. Ibid., November 15, 1833, 199. The Elgin Marbles are Greek sculptures from the Parthenon brought to England at his own expense by Thomas Bruce, 7th Earl of Elgin (ambassador to the Ottoman Empire, 1799–1802), presumably for the sake of preservation. Artists worked for a decade in the early 1800s to remove a number of colossal statues. Several depict naked men on horseback. They were purchased by Parliament in 1816 and subsequently displayed at the British Museum. *The Encyclopedia Americana*, vol. 10 (New York: Americana Corporation, 1953), 238–39.
58. Beagle *Diary*, May 14, 1833, 158.

3. Fossils, Folly & Faults

1. *Charles Darwin's* Beagle *Diary*, October 8, 1832, 109, in *The Complete Work of Charles Darwin Online*, ed. John van Wyhe (2002), http://darwin-online.org.uk/.
2. Malcom Browne, "Legendary Giant Sloth Sought by Scientists in Amazon Rain Forest," *New York Times*, February 8, 1994.

3. *The Princess Bride* (1987), directed by Rob Reiner (Santa Monica: Lionsgate Home Entertainment, 2007), DVD.

4. *Charles Darwin's Zoology Notes and Specimen Lists from H.M.S.* Beagle, ed. Richard D. Keynes (Cambridge: Cambridge University Press, 2000), 152–79.

5. Ibid., 165.

6. Silvia A. Aramay, "A Brief Sketch of the Monte Hermoso Human Footprint Site, South Coast of Buenos Aires Province, Argentina," *Ichnos* 16 (2009): 49–54.

7. Mirta E. Quattrocchio, Cecilia M. Deschamps, Carlos A. Zavala, Silvia C. Grill, and Ana M. Borromei, "Geology of the Area of Bahía Blanca, Darwin's View and the Present Knowledge: A Story of 10 Million Years," *Revista de la Asociación Geológica Argentina* 64, no. 1 (2009): 137–46. Many of the Punta Alta and Monte Hermoso fossils are on display in the Museo Carlos Darwin near Bahía Blanca, where they are interpreted daily for schoolchildren. The Belgrano naval base, home of Argentina's fleet during the Falklands/Malvinas War, now occupies Punta Alta. During the war, a British torpedo sank the *General Belgrano*; 323 Argentine sailors perished. Near the Puerto Belgrano naval station an Argentine billboard beneath the Punta Alta cliffs proudly asserts "Las Malvinas Son Nuestras" (The Malvinas [Falklands] Are Ours).

8. Beagle *Diary*, November 26, 1833, 204.

9. *Darwin's Zoology Notes*, 121.

10. Richard D. Keynes, *Fossils, Finches, and Fuegians: Darwin's Adventures and Discoveries on the* Beagle (New York: Oxford University Press, 2003), 117.

11. Beagle *Diary*, December 17, 1832, 121.

12. Ibid., December 19, 1832, 125, 126.

13. Ibid., December 22–24, 1832, 128.

14. Ibid., December 25–30, 1832, 129–30.

15. Alfred Lansing, Endurance: *Shackleton's Incredible Voyage* (1959; New York: Basic Books, 2007).

16. Dallas Murphy, *Rounding the Horn: Being the Story of Williwaws and Windjammers, Drake, Darwin, Murdered Missionaries and Naked Natives–A Deck's-Eye View of Cape Horn* (New York: Basic Books, 2004), 48.

17. Craig B. Smith, *Extreme Waves* (Joseph Henry Press: Washington, DC, 2006), 76.

18. Beagle *Diary*, January 9, 1832, 131.

19. Ibid., January 11, 1832, 131.

20. Ibid., January 13, 1832, 131.

21. Peter Nichols, *Evolution's Captain: The Story of the Kidnapping That Led to Charles Darwin's Voyage Aboard the Beagle* (New York: Perennial, 2004), 165.

22. Beagle *Diary*, January 13, 1832, 132.

23. Nichols, *Evolution's Captain*, 166.

24. Beagle *Diary*, January 13, 1832, 132.

25. Ibid., January 23, 1833, 138.

26. Ibid., January 19, 1833, 133.

27. Keynes, *Fossils, Finches, and Fuegians*, 127–28.

28. Nichols, *Evolution's Captain*, 177.

29. Beagle *Diary*, January 28, 1833, 139.

30. Ibid.

31. Nichols, *Evolution's Captain*, 188.

32. Beagle *Diary*, January 29, 1833, 140. Keynes includes FitzRoy's account of this incident in a footnote to Darwin's entry. FitzRoy described the boats being "tossed along the beach like empty calabashes." The selection is from Robert FitzRoy, *Narrative of the Surveying Voyages of His Majesty's Ships* Adventure *and* Beagle *Between the Years 1826 and 1836, Describing Their Examination of the Southern Shores of South America, and The*

 Beagle*'s Circumnavigation of the Globe. Proceedings of the Second Expedition, 1831–36, under the command of Captain Robert Fitz-Roy, R.N.* (London: Henry Colburn, 1839), 217, in Complete Work of Charles Darwin Online.

33. Nichols, *Evolution's Captain*, 183.
34. Beagle *Diary*, January 6–15, 1833, 141, 143.
35. Ibid., January 4, 1834, 213.
36. United Press International, "Argentina to Drill for Oil Near Falklands Islands Basin," last modified February 25, 2011, http://www.upi.com/Business_News/Energy-Resources/2011/02/25/Argentina-to-drill-near-Falklands-Islands-basin/UPI-57921298661005/#ixzz1ZwE0hagB (accessed October 6, 2011). The article states that Argentina is pressing ahead with deep-sea drilling plans, financed by Repsol-YPF of Spain and Petrobras of Brazil, "despite risks of heightened tension with Britain."
37. Beagle *Diary*, March 1, 1833, 145.
38. Ibid., March 25, 1833, 148.
39. Ibid., April 16–18, 1833, 151–52.
40. Ibid., December 24, 1833, 208.
41. Ibid., December 25, 1833, 208–9.
42. James Taylor, "Charles Darwin and the *Beagle*," National Archives podcast, May 2009, http://www.nationalarchives.gov.UnitedKingdom/podcasts/darwin-and-beagle.html (accessed October 11, 2011).
43. Beagle *Diary*, December 26, 28; January 1, 3, 1833, 209–11.
44. *Darwin's Zoology Notes*, 186.
45. Ibid., 189.
46. Beagle *Diary*, January 11, 1834, 215. The quotation is from Robert FitzRoy, *Narrative of the Surveying Voyages of His Majesty's Ships* Adventure *and* Beagle (London: Henry Coburn, 1839), 319–20, in van Wyhe, *The Complete Work of Darwin Online*.
47. Charles Darwin, quoted in Keynes, *Fossils, Finches, and Fuegians*, 190.
48. Beagle *Diary*, March 5, 1834, 226.
49. Ibid., 93.
50. Adrian Desmond and James Moore, *Darwin's Sacred Cause: How a Hatred of Slavery Shaped Darwin's Views on Human Evolution* (Boston: Houghton Mifflin Harcourt, 2009).
51. Beagle *Diary*, February 25, 1834, 222–23.
52. Charles Darwin, *The Descent of Man, and Selection in Relation to Sex* (1871), in *From So Simple a Beginning: Darwin's Four Great Books (Voyage of the* Beagle*, The Origin of Species, The Descent of Man, The Expression of Emotions in Man and Animals)*, ed. Edward O. Wilson (New York: W. W. Norton, 2006), 767–1248.
53. Nichols, *Evolution's Captain*, 227, 327.
54. Ibid., 326.
55. Beagle *Diary*, March 6, 1834, 227.
56. Ibid., June 9, 1834, 243.
57. Nichols, *Evolution's Captain*, 203.
58. Beagle *Diary*, June 10, 1834, 244.
59. Ibid., July 8, 1834, 247.
60. Ibid., November 26, 1834, 265.
61. Ibid., December 28, 1834, 276.
62. Ibid., January 19, 1835, 280.
63. Ibid., February 20, 1835, 292.
64. Ibid., March 4, 1835, 295.
65. Ibid., March 5, 1835, 296, 297. Our awareness of the great tsunami tragedy of December 2004, in Indonesia, Sri Lanka, Thailand, and several other Asian nations (the Banda Aceh

quake) and of the horrendous destruction visited upon Japan by the Tōhoku subduction zone earthquake and tsunami of March 2011, provides a clear present-day image of the destruction Darwin must have observed. In May 1960 two great quakes shocked Valdivia and Concepción once again, only this time the center was closer to Valdivia and the energy released thirty times greater. As inferred from historical records in Japan and the evidence of synchronized subsidence along the coasts of Oregon and Washington, a similar- sized earthquake rocked the Pacific Northwest in January 1700. See Brian Atwater, Musumi-Rokkaku Satoko, Satake Kenji, Tsuji Yoshinobu, Ueda Kazue, and David K. Yamaguchi, *The Orphan Tsunami of 1700: Japanese Clues to a Parent Earthquake in North America* (Reston, VA: United States Geological Survey; Seattle, WA: University of Washington Press, 2005).

66. Stephen J. Gould, "Evolution and the Triumph of Homology, or Why History Matters," *American Scientist* 74 (1986): 60–69.

67. Janet Browne, *Charles Darwin: A Biography*, vol. 1, *Voyaging* (Princeton: Princeton University Press, 1995), 531.

4. Irritating Worms

1. Jim Conrad, "Earthworms," *The Backyard Nature Website*, http://www.backyardnature.net/earthwrm.htm (accessed March 27, 2013).

2. *The Autobiography of Charles Darwin 1809–1882*, ed. Nora Barlow (New York: W. W. Norton, 1969), 27.

3. Charles Darwin, *The Formation of Vegetable Mould Through the Action of Worms with Observations on Their Habits* (London: John Murray, 1881), 26–27, in The Complete Work of Charles Darwin Online, ed. John van Wyhe (2002), http://darwin-online.org.uk/.

4. Ibid., 27. As reported in the *New York Times*, September 29, 2015 ("Turning the Worm Around"), neuroscientists at the Salk Institute successfully activated specific neurons in the nematode worm *Caenorhabditis elegans*, using low-pressure ultrasound (low-frequency) signals. The worm responded by making turns after being carefully sensitized. (Wild-type *C. elegans* made no such response.) Stuart Ibsen, Ada Tong, Carolyn Schutt, Sadik Esener, and Sreekanth H. Chalasani, "Sonogenetics Is a Non-Invasive Approach to Activating Neurons in *Caenorhabditis elegans*," *Nature Communications* 6, no. 8264, September 15, 2015, doi:10.1038/ncomms9264.

5. Darwin, *The Formation of Vegetable Mould*, 12.

6. Janet Browne, *Charles Darwin: A Biography*, vol. 2, *The Power of Place* (Princeton: Princeton University Press, 2002), 467, 477.

7. Stephen Jay Gould, "Worm for a Century, and All Seasons," in *Hen's Teeth and Horse's Toes: Further Reflections in Natural History* (New York: W. W. Norton, 1983), 177–86.

8. Darwin, *Formation of Vegetable Mould*, 35.

9. Ibid., 98.

10. Ibid., 97.

11. Doreen Cronin, *Diary of a Worm* (New York: HarperCollins, 2003), April 15 entry, not paginated.

12. Darwin, *Formation of Vegetable Mould*, 76–78.

13. Linley Sambourne, "Man Is But a Worm," *Punch*, December 6, 1881, in *Punch's Almanack for 1882*, http://commons.wikimedia.org/wiki/File:Man_is_But_a_Worm.jpg (accessed March 27, 2013).

14. Browne, *Charles Darwin*, 2:490.

15. Adrian Desmond and James Moore, *Darwin: The Life of a Tormented Evolutionist* (New York: W. W. Norton, 1991), 34.

16. Ibid., 35.

17. Rebecca Stott, *Darwin's Ghosts: The Secret History of Evolution* (New York: Spiegel & Grau, 2012), 88.

18. Ibid.

19. *Charles Darwin's Zoology Notes and Specimen Lists from H.M.S.* Beagle, ed. Richard D. Keynes (Cambridge: Cambridge University Press, 2000), xiii.

20.. Ibid., 195, 201, 202.

21. Ibid., xiv–xv.

22. Desmond and Moore, *Darwin*, 37–38.

23. Robert E. Grant, "Notice Regarding the Ova of the *Pontobdella muricata*, Lam.," *Edinburgh Journal of Science* 7 (1827): 160–61, in van Wyhe, *The Complete Work of Darwin Online.*

24. *Darwin's Zoology Notes*, xvii. Keynes devotes several pages to Darwin's classification of *Flustra* in particular and marine corallines in general, citing "two loose pages of conclusions about the anatomy of the corallines probably written on board the *Beagle* early in 1836" that may be found in the collections of the Cambridge University Library (CUL MS DAR 5.98). Keynes concludes that reading any views on the transmutation of species (descent with modification) into these early notes is unwarranted. Darwin, however, uses the language of "analogy" both in his *Beagle* notes and in the last pages of *Origin.*

25. Charles Darwin, Insectivorous Plants (London: John Murray, 1875), 318, in van Wyhe, *The Complete Work of Darwin Online.*

26. Dewhurst Bilsborrow, "To Erasmus Darwin, On His Work Intitled Zoonomia," in Erasmus Darwin, *Zoonomia; or, The Laws of Organic Life*, vol. 1 (Dublin: B. Dugdale, 1800), vii–viii, http://www.archive.org/stream/zoonomiaorlawsof01darwrich#page/n1/mode/2up (accessed March 17, 2012).

27. Erasmus Darwin, *Zoonomia; or, the Laws of Organic Life*, vol. 1, 2nd American ed., from the 3rd London ed., corrected by the author (Boston: Thomas and Andrews, 1803), 392, reprinted in John H. Wahlert, *Darwin and Darwinism*, http://faculty.baruch.cuny.edu/naturalscience/biology/darwin/biography/erasmus_darwin/zoonomia.html (accessed March 28, 2013). Subsequent references are to this edition.

28. Ibid., 394.

29. Ibid., 398.

30. *Autobiography of Charles Darwin*, 43.

31. Janet Browne, *Charles Darwin: A Biography*, vol. 1, *Voyaging* (Princeton: Princeton University Press, 1995), 133.

32. Desmond and Moore, *Darwin*, 90.

33. *Darwin's Zoology Notes*, xiv.

34. Ibid., 195–99.

35. Charles Darwin, *The Origin of Species by Means of Natural Selection, or The Preservation of Favored Races in the Struggle for Life* (1872), reprint of 6th ed. (Garden City, NY: Doubleday Dolphin Books, [1960), 119.

36. *The Tree of Life Web Project*, http://tolweb.org/tree/ (accessed March 28, 2013).

37. Charles Darwin, *On the Origin of Species by Means of Natural Selection, or The Preservation of Favored Races in the Struggle for Life* (London: John Murray, 1859), 130, in van Wyhe, *The Complete Work of Darwin Online.*

38. Horst Bredekamp, *Darwins Korallen: Frühe Evolutionsmodelle und die Tradition der Naturgeschichte* (Berlin: Klaus Wagenbach Verlag, 2005). There are three synonyms for this alga: *Amphiroa orbigyana*, *Bossea orbigyana*, and *Bossiella orbigyana*. Brederkamp has used *Bossea*. The specimen collected by Darwin and housed in the Natural History Museum, London (Figure 4–1), is

labeled *Amphiroa orbignyana*. http://www.catalogueoflife.org/annual-checklist/2006/show_species_details.php?record_id=18213.

39. Florence Maderspacher, "The Captivating Coral: The Origins of Early Evolutionary Imagery," *Current Biology* 16 (2006): 478, doi 10.1016/j.cub.2006.06.019.

40. Ibid., 476–78.

41. Charles Darwin, "Notebook B: Transmutation of Species," 36–37, transcribed by Kees Rookmaker in van Wyhe, *The Complete Work of Darwin Online.*

42. Ibid., 25.

43. Desmond and Moore, *Darwin*, 40.

44. *Darwin's Zoology Notes*, 228.

45. Desmond and Moore, *Darwin*, 180.

46. *Darwin's Zoology Notes*, xvii.

47. Desmond and Moore, *Darwin*, 136.

48. *Darwin's Zoology Notes*, 232.

49. Darwin, *Formation of Vegetable Mould*, 100.

50. Cronin, *Diary of a Worm*, March 20 entry, not paginated.

51. Darwin, *Formation of Vegetable Mould*, 34, 35.

52. Ibid., 308, 316.

53. Asa Gray, review of *On the Origin of Species*, *Atlantic Monthly* 6 (July–August 1860): 237, in van Wyhe, The Complete Work of Darwin Online.

54. Darwin, *Origin of Species*, 6th ed., 473.

55. Ibid., 473.

56. Ibid., 474.

57. Richard Dawkins, *The Ancestor's Tale: A Pilgrimage to the Dawn of Evolution* (Boston: Houghton Mifflin, 2004), 506–58. Research into the origins and diversification of microbes and protists (single-celled organisms) remains vibrant with many unanswered questions.

58. Darwin, *Formation of Vegetable Mould*, 149–56.

5. A Lungfish Walked into the Zoo

1. Leo Lionni, *Fish Is Fish* (New York: Pantheon Books, 1970), not paginated.

2. Neil H. Shubin, *Your Inner Fish: A Journey into the 3.5 Billion Year History of the Human Body* (New York: Pantheon, 2008).

3. John A. Long, "Dawn of the Deed," *Scientific American* 304, no. 1 (2011): 34–39.

4. Brian K. Hall, "Palaeontology and Evolutionary Developmental Biology: A Science of the Nineteenth and Twenty-First Centuries," *Palaeontology* 45, no. 4 (2002): 647–69.

5. Carol K. Yoon, *Naming Nature: The Clash between Instinct and Science* (New York: W. W. Norton, 2009).

6. Richard Dawkins, *The Ancestor's Tale: A Pilgrimage to the Dawn of Evolution* (Boston: Houghton Mifflin, 2004), summarizes evidence from DNA sequences used to decode the puzzle of common ancestry for all living things. The current status of the "tree of life" solution to this puzzle may be found at The Tree of Life Web Project, http://tolweb.org/tree/.

7. Dawkins, *The Ancestor's Tale*, 54.

8. Christine Janis, personal communication, March 4, 2015.

9. Carl Zimmer, *At the Water's Edge: Fish with Fingers, Whales with Legs, and How Life Came Ashore but Then Went Back to Sea* (New York: Simon & Schuster/Touchstone, 1998), 28–29.

10. Ibid., 100–107.

11. April M. Randal, "Respiratory Behavior and Ecology of the African Air-Breathing Fish *Ctenopoma muriei*," (master's thesis abstract, University of Florida, Gainesville, 2001), http://www.firi.go.ug/Publications/Masters%20Thesis/APRIL_M.pdf (accessed March 19, 2010).

12. Zimmer, *At the Water's Edge*, 101–2.

13. Daniel Goujet, "'Lungs' in Placoderms, a Persistent Palaeobiological Myth Related to Environmental Preconceived Interpretations," *Comptes Rendus Palevol* 10 (2011): 323–29.

14. New World Encyclopedia contributors, "Lungfish," *New World Encyclopedia*, http://www.newworldencyclopedia.org/p/index.php?title=Lungfish&oldid=983858 (accessed January 7, 2016).

15. Jennifer Clack, "From Fins to Limbs," *Natural History* 115 (2006): 36–41.

16. Zimmer, *At the Water's Edge*, 29.

17. Edward B. Daeschler, Neil H. Shubin, and Farish A. Jenkins Jr., "A Devonian Tetrapod-Like Fish and the Evolution of the Tetrapod Body Plan," *Nature* 440, no. 7085 (April 6, 2006): 757–63, doi:10.1038/nature04639; Neil H. Shubin, Edward B. Daeschler, and Farish A. Jenkins Jr., "The Pectoral Fin of *Tiktaalik roseae* and the Origin of the Tetrapod Limb," *Nature* 440, no. 7085 (April 6, 2006): 764–71, doi:10.1038/nature04637.

18. Shubin, *Your Inner Fish*, 25.

19. Sean B. Carroll, *Remarkable Creatures: Epic Adventures in the Search for the Origins of Species* (New York: Houghton Mifflin Harcourt, 2009), 195.

20. Shubin, Daeschler, and Jenkins, "The Pectoral Fin of *Tiktaalik roseae*," 764.

21. Neil H. Shubin, Edward B. Daeschler, and Farish A. Jenkins Jr., "Pelvic Girdle and Fin of *Tiktaalik roseae*," *Proceedings of the National Academy of Sciences* (January 21, 2014): 893–99, doi/10.1073/pnas.1322559111.

22. Chris T. Amemiya, Jessica Alföldi, Alison P. Lee, Fan Shaohua, Philippe Hervé, Iain MacCallum, and Ingo Braasch et al., "The African Coelacanth Genome Provides Insights into Tetrapod Evolution," *Nature* 496, no. 7445 (April 18, 2013): 312, doi:10.1038/nature12027.

23. Ibid., 311.

24. Ibid., 315.

6. Out on a Limb

This chapter is a modestly expanded version of Charles R. Ault Jr., "The Elbow and Ankle Tour of the Zoo," chapter 13 in *Challenging Science Standards: A Skeptical Critique of the Quest for Unity* (Lanham, MD: Rowman & Littlefield, 2016). Reproduced with permission.

1. Stuart Sumida and Beth Rega, "Anatomy and Animations: Combining Classic Techniques with Cutting-Edge Technology," a paper delivered at the National Science Teachers Association Regional Convention, Portland, OR, 2002, provided the inspiration for limb-sketching while visiting zoos, which became an activity in a course for elementary teachers at Lewis & Clark College for many years. Guidance for using zoos as a place for introducing the concepts of homology and cladistics—grouping animals on the basis of shared traits—is provided by Zoos for Effective Science Teaching, a project of the Bronx Zoo Education Department funded by the National Science Foundation (NSF Grant no. TPE 8751488, 1989). Most important, this chapter draws on the experiences of Chuck Jones, in particular his chapter "How to Make a Tennis Shoe for a Percheron," in *Chuck Amuck: The Life and Times of an Animated Cartoonist* (New York: Farrar Straus Giroux, 1989), 252–67. Finally, touring zoos and aquariums through the lens of comparative anatomy

follows the advice in Neil Shubin, *Your Inner Fish: A Journey into the 3.5 Billion Year History of the Human Body* (New York: Pantheon, 2008), 7–10.

2. Jones, *Chuck Amuck*, 258.

3. Ibid.

4. Ibid., 259.

5. Ibid., 260.

6. Ibid., 276.

7. Ibid., 252–53.

8. Chuck Jones tells this story in his chapter "How to Make a Tennis Shoe for a Percheron." As an Oregonian and Portland Trailblazer fan, I have taken license to use the example of a basketball shoe. Jones sketches sneakers that cover the ankle in his drawings of horse, human, kangaroo, and sloth. In his sketches, for all intents and purposes, the shoe is the classic Converse All Star canvas basketball shoe.

9. Ibid., 259.

10 . Kathy Muldoon, "Oregon Zoo Sea Otter Learns to Shoot Hoops," *The Oregonian/Oregon Live*, February 19, 2013, http://www.oregonlive.com/portland/index.ssf/2013/02/oregon_zoo_sea_otter_learns_to.html.

11. Rebecca Morelle, "Elephant's Sixth Toe Discovered," BBC News: Science and Environment, December 22, 2011, http://www.bbc.co.uk/news/science-environment-16250725.

12. Cathy Johnson, *The Sierra Club Guide to Sketching in Nature* (San Francisco: Sierra Club Books, 1997), 37.

13. Claire L. Walker, and Charles E. Roth, *Keeping a Nature Journal: Discover a Whole New Way of Seeing the World around You* (North Adams, MA: Storey, 2003), 111, 181–85.

14. "Natural selection" is not, of course, a conscious agent. There is no "selector"—no wizard behind the screen. Still, in both popular and scientific writing the term has the ring of personification even when none is implied and such may often be the case in my chapters. As Darwin clearly explained, "Everyone knows what is meant and is implied by such metaphorical expressions; and they are almost necessary for brevity. So again it is difficult to avoid personifying the word Nature; but I mean by Nature, only the aggregate action and product of many natural laws, and by laws the sequence of events as ascertained by us. With a little familiarity such superficial objections will be forgotten." Charles Darwin, *The Origin of Species by Means of Natural Selection, or The Preservation of Favored Races in the Struggle for Life* (1872; Garden City, NY: Doubleday Dolphin Books, 1960), 89–90.

15. Joe Raposo, Jon Stone, and Bruce Hart, "One of These Things," *Sesame Street*, PBS (1970).

16. Shubin, *Your Inner Fish*, 7.

7. Nosey Elephants

1. Rudyard Kipling, "The Elephant's Child," in *Just So Stories* (Garden City, NY: Doubleday, 1912), 64–65.

2. Ibid., 71, 65, 76.

3. Ibid., 81.

4. Gregory Bateson, *Mind and Nature* (New York: Bantam, 1988), 16.

5. Christine Janis, personal communication, June 1, 2015.

6. Alfred S. Romer, *The Procession of Life* (Garden City, NY: Anchor Books, 1972), 310.

7. Emmanuel Gheerbrant, "Paleocene Emergence of Elephant Relatives and the Rapid Radiation of African Ungulates," *Proceedings of the National Academy of Sciences Online* 106, no. 26 (2009): 10717–21, doi:10.1073/pnas.0900251106.

8. John P. Hunter and Christine M. Janis, "Spiny Norman in the Garden of Eden? Dispersal and Early Biogeography of Placentalia," *Journal of Mammalian Evolution* 13, no. 2 (2006): 91, doi: 10.1007/s10914–006–9006–6.

9. Ann P. Gaeth, Roger V. Short, and Marilyn B. Renfree, "The Developing Renal, Reproductive, and Respiratory Systems of the African Elephant Suggest an Aquatic Ancestry," *Proceedings of the National Academy of Sciences Online* (1999), doi:10.1073/pnas.96.10.5555.

10. Christine Janis, personal communication, June 1, 2015.

11. Karel Kleisner, Richard Ivell, and Jaroslav Flegr, "The Evolutionary History of Testicular Externalization and the Origin of the Scrotum," *Journal of Bioscience* 35, no. 1 (2010): 27.

12. Christine Janis, personal communication, July 30, 2015.

13. Roger Short, interview by Robyn Williams, "Snorkelling Elephants," Australian Academy of Science (2010), https://www.science.org.au/node/327196#6; Gaeth, Short, and Renfree, "Renal, Reproductive, and Respiratory Systems of the African Elephant."

14. Gaeth, Short, and Renfree, "Renal, Reproductive, and Respiratory Systems of the African Elephant."

15. John B. West, "Why Doesn't the Elephant Have a Pleural Space?" *Physiology* 17, no. 2 (2002): 47–50, http://physiologyonline.physiology.org/content/17/2/.

16. R.E. Brown, J.P. Butler, J.J. Godleski, and S.H. Loring, "The Elephant's Respiratory System: Adaptations to Gravitational Stress," *Respiration Physiology* 109, no. 2 (August 1997): 177–94, doi:10.1016/S0034–5687(97)00038–8.

17. Williams, "Snorkelling Elephants."

18. Ibid.

19. Brown et al., "The Elephant's Respiratory System."

20. Alexander G.S.C. Liu, Erik R. Seiffert, and Elwyn L. Simons, "Stable Isotope Evidence for an Amphibious Phase in Early Proboscidean Evolution," *Proceedings of the National Academy of Sciences Online* (2008): 5786–91, doi:10.1073/pnas.0800884105.

21. Richard Dawkins, *The Ancestor's Tale: A Pilgrimage to the Dawn of Evolution* (Boston: Houghton Mifflin, 2004).

22. Charles Darwin, *The Origin of Species by Means of Natural Selection, or The Preservation of Favored Races in the Struggle for Life* (1872), reprint of 6th ed. (Garden City, NY: Doubleday Dolphin Books, 1960), 76.

23. Ibid., 144.

24. Ibid., 89, 107.

25. Christine Janis, "An Evolutionary History of Browsing and Grazing Ungulates," in *The Ecology of Browsing and Grazing*, ed. Iain J. Gordon and Herbert H.T. Prins (Berlin: Springer-Verlag, 2010), 21.

26. Ibid., 34.

27. Ibid., 41. Janis discusses how gut fermentation places a limit on the amount of time food may remain in passage and the different advantages of foregut and hindgut fermentation. Foregut fermentation in ruminants does have the advantage of detoxifying plant secondary compounds, making otherwise toxic vegetation palatable. Ruminants (extinct and living), however, fail to reach the size of rhinos and elephants. That is because of the limit on how long digesting food may reside in the gut. For the same body size, food passage in ruminants takes longer than in hindgut fermenters (hindgut-fermenting horse: about forty hours; foregut-fermenting cow: approximately sixty hours). Food passage time increases with body size. Passage times that reach four days create a very difficult metabolic problem, leading to high losses of energy as methanogenic bacteria convert acetic acid from the fermentation of cellulose to methane and carbon dioxide. "Elephants solve this problem

by adopting relatively shorter and broader guts, thus speeding up the passage rate of the digesta; but this would not be a possible solution for a ruminant, where the rumenoreticulum portion of the stomach is specifically adapted to delay food passage. Thus ruminants are limited to body sizes where their digesta passage time is less than four days (likely under 1,500 kg)" (ibid.).

28. Jeheskel Shoshani, "Understanding Proboscidean Evolution: A Formidable Task," *Trends in Ecology & Evolution* 13, no. 12 (1998): 480.

29. Janis, "Evolutionary History of Browsing and Grazing Ungulates," 31.

30. Charles Darwin, *On the Origin of Species by Means of Natural Selection, Or the Preservation of Favoured Races in the Struggle for Life*, 1st ed. (London: Murray, 1859), 82, in *The Complete Work of Darwin Online*, ed. John van Wyhe (2002), http://darwin-online.org.uk/.

31. Curtis Johnson, *Darwin's Dice: The Idea of Chance in the Thought of Charles Darwin* (Oxford: Oxford University Press, 2015), xxiii.

32. Darwin, *Origin of Species*, 1st ed., 82.

33. Ibid.

34. Yolanda Pretorius, Willem F. de Boer, Kim Kortekaas, Machiel van Wijngaarden, Rina C. Grant, Edward M. Kohi, Emmanuel Mwakiwa, Rob Slotow, and Herbert H.T. Prins, "Why Elephant Have Trunks and Giraffe Long Tongues: How Plants Shape Large Herbivore Mouth Morphology," *Acta Zoologica* (Stockholm, February 2015): 6–7, doi:10.1111/azo.12121.

35. Ibid., 7.

36. Jeheskel Shoshani and Pascal Tassy, "Advances in Proboscidean Taxonomy & Classification, Anatomy & Physiology, and Ecology & Behavior," *Quaternary International* 126–128 (2005): 14.

37. Jeheskel Shoshani, "Elephant: Sound Production and Water Storage," *Encyclopaedia Britiannica*, http://www.britannica.com/animal/elephant-mammal#ref800402 (accessed January 11, 2016).

38. Shoshani and Tassy, "Advances in Proboscidean Taxonomy & Classification, Anatomy & Physiology, and Ecology & Behavior," 14.

39. Douglas J. Emlen, *Animal Weapons: The Evolution of Battle* (New York: Henry Holt), 60–63, quote 61.

40. Ibid, 62.

41. Ibid., 74.

42. Elephant Voices Contributors, "Elephant Communication: Acoustic Communication/Sound Detection," *Elephant Voices*, http://www.elephantvoices.org/elephant-communication/acoustic-communication.html (accessed January 8, 2016).

43. Shoshani, "Understanding Proboscidean Evolution," 483.

44. Ibid., 484.

45. Ibid., 482.

46. Ibid., 485.

47. Benjamin L. Hart, Lynette A. Hart, and C. R. Sarath, "Cognitive Behavior in Asian Elephants: Use and Modification of Branches for Fly Switching," *Animal Behaviour* 62, no. 5 (2001): 839, doi:10.1006/anbe.2001.1815.

48. Lucy A. Bates, Joyce H. Poole, and Richard W. Byrne, "Elephant Cognition," *Current Biology* 18, no. 13 (2008): R544, doi:10.1016/j.cub.2008.04.019.

49. Theodor Geisel (Dr. Seuss), Horton Hatches the Egg (1940; New York: Random House, 1968).

50. Stephan Klassen, "The Relation of Story Structure to a Model of Conceptual Change in Science Learning," *Science & Education* 19 (2010): 314, doi:10.1007/s11191–009–9212–8.

8. The Bearduck of Baleen

1. Herman Melville, *Moby-Dick* (1851; New York: Bobbs-Merrill, 1964), 193.

2. Rudyard Kipling, "How the Whale Got His Throat," in *Just So Stories* (Garden City, NY: Doubleday, 1912), 3, 7, 11.

3. Nathaniel Philbrick, *In the Heart of the Sea: The Tragedy of the Whaleship* Essex (New York: Viking, 2000).

4. Charles Darwin, *The Origin of Species by Means of Natural Selection, Or the Preservation of Favoured Races in the Struggle for Life*, 6th ed. (London: Murray, 1872), 183, in *The Complete Work of Charles Darwin Online*, ed. John van Wyhe (2002), http://darwin-online.org.uk/.

5. Ibid.

6. Ibid.

7. Charles Darwin, *On the Origin of Species by Means of Natural Selection. Or the Preservation of Favoured Races in the Struggle for Life*, 1st ed. (London: Murray, 1859), 184, in van Wyhe, *The Complete Work of Darwin Online.*

8. John Morris, "Review of *On the Origin of Species*," *Dublin Review* 48 (1860): 64, in van Wyhe, *The Complete Work of Darwin Online.*

9. R. B. Freeman, "Whale-Bear Story," *Charles Darwin: A Companion: 2nd Online Edition* (2007), comp. Sue Asscher, in van Wyhe, *The Complete Work of Darwin Online.*

10. Philip D. Gingerich, Munir ul Haq, Iyad S. Zalmout, Intizar Hussain Khan, and M. Sadiq Malkani, "Origin of Whales from Early Artiodactyls: Hands and Feet of Eocene Protocetidae from Pakistan," *Science* 293, no. 5538 (2001): 2239–42.

11. Adrian Desmond and James Moore, *Darwin: The Life of a Tormented Evolutionist* (New York: W. W. Norton, 1991), 570. The joint paper by Darwin and Wallace, "On the Tendency of Species to Form Varieties; and on the Perpetuation of Varieties and Species by Natural Means of Selection," was read on July 1, 1858, to the fellows of the Linnean Society in London. *On the Origin of Species* by Darwin alone went to press the following year.

12. Ernst Mayr, *What Evolution Is: From Theory to Fact* (New York: Basic Books, 2001), 205.

13. Sean B. Carroll, *The Making of the Fittest: DNA and the Ultimate Forensic Record of Evolution* (New York: W.W. Norton, 2006), 205–7.

14. Edward Humes, *Monkey Girl: Evolution, Education, Religion, and the Battle for America's Soul* (New York: HarperCollins, 2007).

15. Robert W. Meredith, John Gatesy, Joyce Cheng, and Mark Springer, "Pseudogenization of the Tooth Gene Enamelysin (MMP20) in the Common Ancestor of Extant Baleen Whales," *Proceedings of the Royal Society of Biological Sciences* 278 (2010): 993–1002, doi:10.1098/rspb.2010.1280.

16. Thomas A. Deméré and Annalisa Berta, "Skull Anatomy of The Oligocene Toothed Mysticete *Aetioceus* [*sic*] *Weltoni* (Mammalia; Cetacea): Implications For Mysticete Evolution And Functional Anatomy," *Zoological Journal of the Linnean Society* 154 (2008): 308–52, doi: 10.1111/j.1096-3642.2008.00414.x.

17. Charles B. Clayman, *The American Medical Association Encyclopedia of Medicine* (New York: Random House, 1989), 614.

9. The Saga of Mooshmael

1. All quotations are from Bernard Wiseman, *Morris the Moose* (1959; New York: HarperCollins 1989).

2. Gareth B. Matthews, *Philosophy and the Young Child* (Cambridge: Harvard University Press, 1980), 95.

3. Associated Press. "Horse, Moose Become Friends in Vermont," *USA Today*, August 30, 2004, http://www.usatoday.com/news/nation/2004–08–30-moose-cow_x.htm; "Bull-Headed," *Indianapolis Star*, November 9, 1986.

4. Pat A. Wakefield and Larry Carrara, *A Moose for Jessica* (New York: E.P. Dutton, 1987).

5. Matthews, *Philosophy and the Young Child*, 76–77.

6. *Pakicetus* and *Indohyus*, the extinct beasts among the ancestors of modern cetaceans, are considered transitional between terrestrial creatures and whales, a classification supported by similar anatomy of the inner ear. This evolution is described in greater and more technical detail in the next chapter, "The Higgledy-Piggledy Whale." See also Philip D. Gingerich, B. Holly. Smith, and Elwyn. L. Simons, "Hind Limbs of Eocene *Basilosaurus isis:* Evidence of Feet in Whales," *Science* 249, no 4965 (1990): 154–57.

7. Herman Melville, *Moby-Dick or The Whale* (1851; New York: Bobbs-Merrill, 1964), 448.

8. J. G. M. Thewissen, Lisa Noelle Cooper, Mark T. Clementz, Sunil Bajpai, B. N. Tiwari et al., "Whales Originated from Aquatic Artiodactyls in the Eocene Epoch of India," *Nature* 450 (2007): 1190–95, doi:10.1038/nature06343.

9. Among the Artiodactyla the Hippopodimidae claim closest affinity to the Cetacea. Doris's family album may have introduced Mooshmael to her pedigree as a Whippomorph (whales and hippos—and extinct hippish whales/whaley hippos). It is likely that she shared some stories about her hippopodimid cousins and their long-lasting embrace of an aquatic lifestyle. Mooshmael is currently planning an African safari in hopes of meeting his Whippomorph relatives. For the introduction of the term "Whippomorpha" (a clade joining whales to hippos) see Peter J. Waddell, Norihiro Okada, and Masami Hasegawa, "Towards Resolving the Interordinal Relationships of Placental Mammals," Systematic Biology 48 (1999): 1–5, doi:10.1093/sysbio/48.1.1. JSTOR2585262 PMID12078634.

10. The Higgledy-Piggledy Whale

1. Lisa Noelle Cooper, J. G. M. Thewissen, Sunil Bajpai, and B. N. Tiwari, "Postcranial Morphology and Locomotion of the Eocene Raoellid Indohyus (Artiodactyla: Mammalia)," *Historical Biology* 24, no. 3 (2012): 279–310, doi:10.1080/08912963.2011.624184.

2. Philip D. Gingerich, "Evidence for Evolution in the Vertebrate Fossil Record," *Journal of Geological Education* 31, no. 2 (1983): 143.

3. John A. McPhee, *Coming into the Country* (New York: Farrar, Straus and Giroux, 1977).

4. Gingerich, "Evidence for Evolution," 143.

5. Ibid., 144.

6. Xiaoyuan Zhou, Rhenji Zhai, Philip D. Gingerich, and Liezu Chen, "Skull of a New Mesonychid (Mammalia, Mesonychia) from the Late Paleocene of China," *Journal of Vertebrate Paleontology* 15, no. 2 (1995): 387–400.

7. J. G. M. Thewissen. *The Walking Whales: From Land to Water in Eight Million Years* (Oakland, CA: University of Chicago Press, 2014), 128. In more technical terms, "double-pulley" refers to "double-trochleated," meaning the rounded hinge shape at both ends of the astragalus. As explained by Thewissen, "All artiodactyls are characterized by the particular shape of a bone in the ankle, the astragalus. In all mammals, the astragalus is the bone on which the ankle pivots. To allow that, the bone has a hinge joint called the trochlea that articulates with the shine bone (tibia) above it. The other side of the astragalus is an area called the head. It faces the foot and has different shapes in different mammals. It is globular in most mammals, flat in horses, and has the shape of another trochlea in artiodactyls. This

double-trochleated astragalus is very distinctive, characterizing all artiodactyls from the smallest mouse deer to the larges giraffe, including all the fossil ones."

8. Philip D. Gingerich, B. Holly Smith, and Elwyn L. Simons, "Hind Limbs of Eocene *Basilosaurus isis:* Evidence of Feet in Whales," *Science* 249, no. 4965 (1990): 154–57.

9. Peter J. Waddell, Norihiro Okada, and Masami Hasegawa, "Towards Resolving the Interordinal Relationships of Placental Mammals," Systematic Biology 48 (1999): 1–5, doi:10.1093/sysbio/48.1.1. JSTOR 2585262. PMID 12078634.

10. Jonathan H. Geisler and Jessica M. Theodor, "Hippopotamus and Whale Phylogeny," *Nature* 458, no. 7236 (2009): E1–E4, doi:10.1038/nature07776.

11. Kate Wong, "The Mammals That Conquered the Sea," *Scientific American* 286, no. 5 (2002): 72–73.

12. Steve Jones, *Darwin's Ghost: The Origin of Species Updated* (New York: Random House, 2000), 18–19.

13. Geisler and Theodor, "Hippopotamus and Whale Phylogeny," E1. Geisler and Theodor draw two phylogenetic trees positioning the cetaceans and hippopotomids in relation to the artiodactyls and mesonychids. Their data favor an affinity between ruminants (e.g., deer) and whale-hippo branches. Previous thinking associated ruminants more closely with camels and pigs, leaving hippos with whale connections farther removed (and mesonychids more distant still).

14. J. Gatesy, C. Hayashi, M. A. Cronin, and P. Arctander, "Evidence from Milk Casein Genes That Cetaceans Are Close Relatives of Hippopotamid Artiodactyls," *Molecular Biology and Evolution* 13, no. 7 (1996): 954–63.

15. Graham D. Burnett, *Trying Leviathan: The Nineteenth-Century Court Case That Put the Whale on Trial and Challenged the Order of Nature* (Princeton: Princeton University Press, 2007).

16. Ibid., 4. Burnett drew heavily on the historical account of the *Maurice v. Judd* trial as recorded in William Sampson, *Is a Whale a Fish? An Accurate Report of the Case of James Maurice against Samuel Judd* (New York: Van Winkle, 1818).

17. Herman Melville, *Moby-Dick or The Whale* (1851; New York: Bobbs-Merrill, 1964), 182.

18. Ibid. 183.

19. Ibid.

20. Gregory Bateson, *Mind and Nature: A Necessary Unity* (New York: Bantam, 1988), 16.

21. Melville, *Moby-Dick*, 436.

22. Ibid., 446.

23. J. G. M. Thewissen, Lisa Noelle Cooper, M. T. Clementz, and Sunil Bajpai, "Whales Originated from Aquatic Artiodactyls in the Eocene Epoch of India," *Nature* 450, no. 7173 (2007): 1190–95, doi:10.1038/nature06343.

24. J. G. M. Thewissen, E.M. Williams, L. J. Roe, and S. T. Hussain, "Skeletons of Terrestrial Cetaceans and the Relationship of Whales to Artiodactyls," *Nature* 413, no. 6853 (2001): 277–81.

25. J. G. M. Thewissen, S. Taseer Hussain, and M. Arif, "Fossil Evidence for the Origin of Aquatic Locomotion in Archaeocete Whales," *Science* 263 (1994): 210–12.

26. Phillip D. Gingerich, Munir ul Haq, Iyad S. Zalmout, Intizar Hussain Khan, and M. Sadiq Malkani, "Origin of Whales from Early Artiodactyls: Hands and Feet of Eocene Protocetidae from Pakistan," *Science* 293, no. 5538 (2001): 2240.

27. Ibid., 2241.

28. Gingerich et al., "Origin of Whales from Early Artiodactyls," 2239–42.

29. Philip D. Gingerich, "Evolution of Whales from Land to Sea," *Proceedings of the American Philosophical Society* 156, no. 3 (2012): 309–23.

30. Wong, "The Mammals That Conquered the Sea," 76.

31. Thewissen, Cooper, Clementz, and Bajpai, "Whales Originated from Aquatic Artiodactyls in the Eocene Epoch of India," 1190.

32. J. G. M. Thewissen and Sunil Bajpai, "Whale Origins As a Poster Child for Macroevolution," *BioScience* 51, no. 12 (2001): 1037–49, http://www.jstor.org/stable/10.1641/0006–3568%282001%29051%5B1037%3AWOAAPC%5D2.0.CO%3B2.

33. Wong, "The Mammals That Conquered the Sea," 77.

34. Jonathan H. Geisler, Matthew W. Colbert, and James L. Carew, "A New Fossil Species Supports an Early Origin for Toothed Whale Echolocation" *Nature* 508, no. 7496 (2014): 383–86, doi:10.1038/nature13086.

35. Wong, "The Mammals That Conquered the Sea," 77.

36. Robert W. Meredith et al., "Pseudogenization of the Tooth Gene Enamelysin (MMP20) in the Common Ancestor of Extant Baleen Whales," *Proceedings of the Royal Society of Biological Sciences* 278 (2010): 993–1002, doi:10.1098/rspb.2010.1280.

37. Thomas A. Deméré, M. R. McGowen, A. Berta, and J. Gatesy, "Morphological and Molecular Evidence for a Stepwise Evolutionary Transition from Teeth to Baleen in Mysticete Whales," *Systematic Biology* 57, no. 1 (2008): 15–37, doi:10.1080/10635150701884632.

11. Archaic Chickengators

1. Leo Lionni, *An Extraordinary Egg* (New York: Random House, 1994), 27, 28.

2. Lindsay E. Zanno, Susan Drymala, Sterling J. Nesbitt, and Vincent P. Schneider, "Early Crocodylomorph Increases Top Tier Predator Diversity during Rise of Dinosaurs," *Scientific Reports* 5 (2015): 9276, doi:10.1038/srep09276.

3. Thomas R. Holtz Jr., *Dinosaurs: The Most Complete, Up-to-Date Encyclopedia for Dinosaur Lovers of All Ages*, illus. Luis V. Rey (New York: Random House, 2007), 57, 62.

4. British Museum of Natural History, *Dinosaurs and Their Living Relatives* (Cambridge: British Museum of Natural History and the Press Syndics of the University of Cambridge, 1985).

5. Alfred S. Romer, *The Vertebrate Body* (Philadelphia: W.B. Saunders, 1962), 231–32.

6. Daniel J. Field, Jacques A. Gauthier, Benjamin L. King, Davide Pisani, Tyler R. Lyson, and Kevin J. Peterson, "Toward Consilience in Reptile Phylogeny: miRNSs Support an Archosaur, Not Lepidosaur, Affinity for Turtles," *Evolution & Development* 16, no. 4 (2014): 189–96, doi:10.1111/ede.12081; Ylenia Chiari, Vincent Cahais, Nicolas Galtier, and Frédéric Delsuc, "Phylogenomic Analyses Support the Position of Turtles as the Sister Group of Birds and Crocodiles (Archosauria))," *BioMed Central Biology* 10, no. 65 (July 27, 2012): n.p., doi:10.1186/1741–7007–10–65.

7. Gregory M. Erickson, Paul M. Gignac, Scott J. Steppan, A. Kristopher Lappin, Kent A. Vliet, John D. Brueggen, Brian D. Inouye, David Kledzik, and Grahame J.W. Webb, "Insights into the Ecology and Evolutionary Success of Crocodilians Revealed through Bite-Force and Tooth-Pressure Experimentation," *Plos One* 7, no. 3 (2012): 5, 10 e31781, doi:10.1371/journal.pone.00317815.

8. Holtz, *Dinosaurs*.

9. Richard O. Prum and Alan H. Brush, "Which Came First, the Feather or the Bird?" *Scientific American Special Edition* 14, no. 2 (Summer 2014): 80.

10. Ibid., 82.

11. Ibid., 84.

12. British Museum of Natural History, *Dinosaurs and Their Living Relatives*.

13. John H. Ostrom, "Archaeopteryx and the Origin of Birds," *Biological Journal of the Linnean Society* 8 (1976): 91–182.

14. Holtz, *Dinosaurs*, 166.

15. Ibid.

16. Ibid., passim.

17. Xing Xu, Kebai Wang, Ke Zhang, Qingyu Ma, Lida Xing, Corwin Sullivan, Dongyu Hu, Shuqing Cheng, and Shuo Wang, "A Gigantic Feathered Dinosaur from the Lower Cretaceous of China," *Nature* 484 (2012): 92–95, doi:10.1038/nature10906.

18. Holtz, *Dinosaurs*, 166–67.

19. Stephen M. Gatesy and Kenneth P. Dial, "From Frond to Fan: Archaeopteryx and the Evolution of Short-Tailed Birds," *Evolution* 50, no. 5 (1996): 2037–48, http://www.jstor.org/stable/2410761 (accessed March 7, 2014).

20. Prum and Brush, "Which Came First, the Feather or the Bird?"; Xiao-Ting Zheng, Hai-Lu You, Xing Xu, and Zhi-Ming Dong, "An Early Cretaceous Heterodontosaurid Dinosaur with Filamentous Integumentary Structures," *Nature* 458 (2009): 333–36.

21. Stephen Jay Gould, *Hen's Teeth and Horse's Toes: Further Reflections in Natural History* (New York: Norton, 1983); Ed Kollar and Chris Fisher, "Tooth Induction in the Chick Epithelium: Expression of Quiescent Genes for Enamel Synthesis," *Science* 207 (1980): 993–95.

22. Maiko Watanabe, Masato Nikaido, Tomi T. Tsuda, Takanori Kobayashi, and David Mindell, "New Candidate Species Most Closely Related to Penguins," *Gene* 378 (2006): 65–73, doi:10.1016/j.gene.2006.05.003.

23. Erich D. Jarvis, Siavash Mirarab, Andre J. Aberer, Bo Li, Peter Houde, Cai Li, Simon Y.W. Ho, et al., "Whole-Genome Analyses Resolve Early Branches in the Tree of Life of Modern Birds," Science 346, no. 6215 (2014): 1320–31, doi:10.1126/science.1253451.

24. Lianhai Hou, Larry D. Martin, Zhonghe Zhou, Alan Feduccia, and Fucheng Zhang, "A Diapsid Skull in a New Species of the Primitive Bird *Confuciusornis*," *Nature* 399 (1999): 679–82.

25. David E. Alexander, Enpu Gong, Larry D. Martin, and David A. Burnham, "Model Tests of Gliding with Different Hindwing Configurations in the Four-Winged Dromaeosaurid *Microraptor gui*," *Proceedings of the National Academy of Sciences* 107, no. 7 (February 16, 2010): 2972–76, doi/10.1073/pnas.0911852107.

26. Quanguo Li, K-Qin Gao, Qingjin Meng, Julia A. Clarke, Matthew D. Shawkey, Liliana D. Alba, Rui Pei, Mick Ellison, Mark A. Norell, and Jakob Vinther, "Reconstruction of Microraptor and the Evolution of Iridescent Plumage," *Science* 335, no. 6073 (2012): 1215–19, doi: 10.1126/science.1213780.

12. Coral Pigs and Tide Pool Sheep

1. Bill Peet, *Chester the Worldly Pig* (New York: Houghton Mifflin, 1965), 8.

2. Thomas J. O'Shea, "Manatees," *Scientific American* 271 (1994): 66.

3. Charles Darwin, *The Origin of Species by Means of Natural Selection, or The Preservation of Favored Races in the Struggle for Life* (1872; Garden City, NY: Doubleday Dolphin Books, 1960), 173.

4. Ibid.

5. Ibid., 174.

6. Ibid., 89–90.

7. Ibid., 95–97.

8. Ibid., 95.

9. Ernst Mayr, *What Evolution Is: From Theory to Fact* (New York: Basic Books, 2001), 186–87.

10. Ibid., 181.

11. Charles Darwin, *The Voyage of the* Beagle*: Darwin's Five-Year Circumnavigation* (Santa Barbara, CA: Narrative Press, 2001), 54.

12. Ibid., 55–56.

13. Curtis Johnson, *Darwin's Dice: The Idea of Chance in the Thought of Charles Darwin* (Oxford: Oxford University Press, 2015), 11–21.

14. Adrian Desmond and James Moore, *Darwin: The Life of a Tormented Evolutionist* (New York: W. W. Norton, 1991), 369.

15. Muhammad Ali Khalidi, "Carving Nature at the Joints," *Philosophy of Science* 60, no. 1 (1993):100, http://www.jstor.org/stable/188457.

16. Philip Kitcher, *The Advancement of Science: Science without Legend, Objectivity without Illusions* (New York: Oxford University Press, 1993).

Epilogue

Elements of this epilogue first appeared in Toby R. Ault and Charles R. Ault Jr., "On the Trail of Darwin's Megabeasts," *American Paleontologist* 17, no. 1 (2009): 16–19. Reprinted with permission from the Paleontological Research Institution, Ithaca, NY.

1. Martin Lockley, *Tracking Dinosaurs: A New Look at an Ancient World* (New York: Cambridge University Press, 2006), 128.

2. India Wood, personal correspondence, February 12, 2015. India is working on a memoir that will include her *Allosaurus* discovery.

3. India Wood, personal correspondence, September 6, 2014.

4. Kirk Johnson, *Cruisin' the Fossil Freeway* (Golden, CO: Fulcrum Publishing, 2007), 139. In Johnson's telling of the story of "India Wood's *Allosaurus* Skeleton," India called the museum and said, "I've got an Allosaurus in my bedroom, do you want to see it?" India clarified in her personal correspondence that she actually wrote a note and received a phone call from the museum. Johnson's version of India's first contact with the museum makes a better story.

5. Wood, personal correspondence, September 6, 2014.

6. Juan Carlos Fernicola, Sergio F. Vizcaino, and Gerardo de Iuliis, "The Fossil Mammals Collected by Charles Darwin in South American during His Travels on Board the HMS *Beagle*," *Revista de la Asociación Geológica Argentina* 64, no. 1 (2009): 147–59, http://www.scielo.org.ar/pdf/raga/v64n1/v64n1a16.pdf.

7. Charles Darwin, *The Voyage of the* Beagle*: Darwin's Five-Year Circumnavigation* (1845; Santa Barbara: Narrative Press, 2001), 74.

8. *Charles Darwin's* Beagle *Diary*, ed. Richard D. Keynes (Cambridge: Cambridge University Press, 2001), October 8, 1832, 109.

9. Darwin, *Voyage of the* Beagle, 183.

10. Ault and Ault, "On the Trail of Darwin's Megabeasts."

11. Silvia A. Aramayo, "A Brief Sketch of the Monte Hermoso Human Footprint Site, South Coast of Buenos Aires Province, Argentina," *Ichnos* 16 (2009): 49–54.

12. Teresa Manera de Bianco and Silvia A. Aramayo, "Yacimiento Paleoicnológico de Pehuen-Có: Guia de Campo," paper presented at the XVIII Jornadas Argentinas de Paleotología de Vertebrados, Bahía Blanca, Argentina, May 25, 2002.

13. Darwin, *Voyage of the* Beagle, 182.

14. Ibid., 87.

15. Richard D. Keynes, ed., *Charles Darwin's* Beagle *Diary* (Cambridge: Cambridge University Press, 2001), May 9, 1833, 155.

16. Charles R. Ault Jr., "Intelligently Wrong: Some Comments on Children's Misconceptions," *Science & Children* 21, no. 8 (1984) : 22.

17. Charles Darwin, *Journal of Researches into the Natural History and Geology of the Countries Visited During the Voyage of H.M.S.* Beagle *Round the World, under the Command of Capt. Fitz Roy, R.N.*, 2nd ed. (London: John Murray, 1845), 173, in *The Complete Work of Charles Darwin Online*, ed. John van Wyhe (2002), http://darwin-online.org.uk/.

INDEX

Page locators in italics refer to figures.